AMERICAN NATIONAL STANDARD
ENGINEERING DRAWING AND RELATED
DOCUMENTATION PRACTICES

Dimensioning and Tolerancing

ANSI Y14.5 - 1973

For uniformity, all the linear dimensions in this American National Standard are shown in U.S. customary (inch) units (exception 5-7). It should be understood that SI (metric) units of measurement could equally well have been used without prejudice to the principles established herein. See 5-1.8.

SECRETARIAT

AMERICAN SOCIETY OF ENGINEERING EDUCATION
THE AMERICAN SOCIETY OF MECHANICAL ENGINEERS
SOCIETY OF AUTOMOTIVE ENGINEERS

PUBLISHED BY

THE AMERICAN SOCIETY OF MECHANICAL ENGINEERS

United Engineering Center   345 East 47th Street   New York, N. Y. 10017

FOREWORD

This Foreword is not a part of American National Standard for Dimensioning
and Tolerancing, Y14.5-1973

This issue is a revision of American National Standard Y14.5-1966, Dimensioning and Tolerancing for Engineering Drawings. Additional international practices have been introduced in order to facilitate a common drawing interpretation for those products designed in the United States and subject to world-wide manufacturing and distribution.

Work on this issue began with a meeting in Albuquerque, in October, 1967. Decisions were reached regarding subjects planned for introduction as well as proposed improvements and refinements to information already published.

A series of meetings was held during 1968 through 1970. Early meetings involved the discussion of technical subjects for which explanatory papers were written and carefully analyzed. Following these discussions, the proposed text and illustrations were prepared and reproduced in a succession of drafts. At each stage of development, comments were introduced, reviewed by the Subcommittee and resolved by consensus. Finally, a preliminary draft evolved which was submitted for review and approval of the Subcommittee in December 1971.

During the development of this issue close liaison was maintained with the British Standards Institution and the Canadian Standards Association. British and Canadian representatives attended many Y14.5 meetings, as parallel efforts were underway to revise their corresponding national standards, BS308 and CSA B78.2. Constructive comments were exchanged which proved to be of mutual benefit.

Several international meetings on dimensioning and tolerancing occurred since 1966 which influenced the direction of change and broadened the base about which this standard was developed. Participation by members of the Y14.5 Subcommittee in American-British-Canadian (ABC) conferences provided the opportunity for unifying the dimensioning and tolerancing practices of the major English-speaking countries. Y14.5 representation on the U.S. Committee for ISO/TC 10, Technical Committee on Engineering Drawing Principles, of the International Organization for Standardization (ISO) has ensured a commonality in basic dimensioning and tolerancing principles as recognized on a world-wide basis.

At the Moscow TC 10 meeting on drawing practices in June, 1967, the United States delegation introduced papers for consideration on "The Three-Plane Concept" and "The Projected Tolerance Zone Concept," both concepts already featured in the Y14.5-1966 Standard.

An ABC conference followed in Ottawa, in May, 1968. Thirteen U.S. position papers were submitted on principal differences between the three national standards. Agreements were reached between the parties on unifying certain practices in order to narrow existing differences. It afforded the opportunity to better understand each other's viewpoint and reasoning, which is fundamental in the achievement of ABC unification.

Another ABC conference followed in December, 1970, in London. As a basis for conference discussion, the United States delegation presented nine papers. A reinforcement of agreements reached at Ottawa was in evidence when proposed changes to the national standards of the three countries were reviewed. The London meeting provided agreement to continue the effort of ABC unification in dimensioning and tolerancing and jointly pursue investigation of new concepts and practices.

The advances taken by this standard toward international practices are of importance because of the increasing United States participation in world trade and commerce. With the publication of this issue of the standard, the desire for a common interpretation of dimensioning and tolerancing practices, regardless of origin, moves closer to practical implementation.

Members of the Y14.5 Subcommittee who participated in this revision and provided the technical expertise represent a broad cross section of American industry, the Department of Defense (D.O.D.), and educational institutions. Liaison with technical societies such as American Society of Mechanical Engineers (ASME), Society of Automotive Engineers (SAE), American Ordnance Association (AOA) and Aerospace Industries Association (AIA) provided additional technical supplementation. Particular recognition is given to the industries, DOD agencies, and educational institutions who found it possible to sponsor the participants in these activities. The success of this effort can be attributed to their demonstrated interest, cooperation, and strong support.

The following is a summary of the significant changes and improvements made in this issue of American National Standard Y14.5:

- The definition of true position is given. Also, the international expression "positional tolerancing" is adopted as a standard term of reference in place of "true position tolerancing."(See 5-1.6.3)

- The definition of a feature is clarified.(See 5-1.6.7)

- The expression "least material condition" and its abbreviation LMC is recognized as a standard term of reference.(See 5-1.6.12)

- Virtual condition (sometimes referred to as virtual size) is defined.(See 5-1.6.14)

- The international expression "full indicator movement" and its abbreviation FIM is adopted as a standard term of reference.(See 5-1.6.20)

- The international practices of enclosing a reference dimension with parentheses and underlining a "not-to-scale" dimension with a straight line are introduced. (See 5-1.10.5 and 5-1.10.8)

- A footnote indicates where the Department of Defense takes exception to the stated requirement. (See 5-2.1.1 and 5-2.4.1)

- The method of expressing limit dimensions is standardized to agree with international practice.(See 5-2.2)

- The practice of always specifying whether MMC or RFS applies to an individual tolerance, datum reference, or both is established as the preferred method for positional tolerancing. A cautionary note is given for the former method, now considered an alternate practice.(See 5-2.12)

- Explanation of the effects of RFS, MMC, and zero tolerance at MMC is improved to facilitate comparison.(See 5-2.12.1, 5-2.12.2 and 5-2.12.3)

- An explanation of the virtual condition boundary is provided. (See 5-2.15)

- A general rule is introduced for datum features of size that, in addition to being datums, have a separate tolerance of position or form applied to them.(See 5-2.15.1)

- Geometric characteristic symbols are appropriately recategorized.(See Figure 73)

- The slanted parallelism symbol is adopted and a diameter symbol is introduced. Both symbols are recognized internationally. (See Figure 73 and 5-3.3.5)

- New symbols are introduced to indicate a projected tolerance zone and a datum target.(See 5-3.3.6 and 5-3.3.8)

- The international sequence of symbols within a feature control symbol is introduced as a permissible practice.(See 5-3.4.2)

- A composite feature control symbol for positional tolerancing applications is introduced. It is to be used for specifying dual requirements for the same group of features.(See 5-3.4.2.2)

- Datum referencing (explained in former Appendix A, 1966 issue) is incorporated in this standard. Datum target practices are standardized.(See 5-4)

- The use of specified datums for positional tolerancing examples is amplified.(See 5-5)

- The explanation of separate requirements for multiple patterns of features relative to common datums is expanded. (See 5-5.6.3)

- The symbol method for expressing tolerances of position and form is given predominant recognition. (See 5-5 and 5-6)

- The provision for specifying a straightness tolerance on an MMC or RFS basis is reestablished. A straightness tolerance applied in two directions to the line elements of a flat surface is introduced. (See 5-6.4.1.2 and 5-6.4.1.4)

- The explanation of profile tolerancing has been expanded. New items featured are: application of datums; the use of profile tolerances in combination with other form tolerances; and profile tolerances applied to coplanar surfaces. (See 5-6.5.4, 5-6.5.5, and 5-6.5.6)

- An example is given for specifying an angularity tolerance for an axis. (See Figure 161)

- The method of specifying runout tolerances is revised. Circular runout (recognized as runout in international standards) is specified by the use of the runout symbol without indicating the word CIRCULAR. Total runout, formerly identified as runout in the previous issue, is specified by the use of the runout symbol, supplemented by the word TOTAL. (See 5-6.7.2)

- Dual dimensioning is introduced as a method for specifying both inch and metric units on a single drawing. (See 5-7)

- Information on dimensioning for numerical control is introduced in Appendix A.

- Information on zero positional tolerancing at MMC is included in Appendix B.

- Information on the form and proportion of individual geometric characteristic symbols is introduced in Appendix D.

A draft of this issue was released for industry and DOD review in January 1972. All comments received were acknowledged, evaluated and the authors of comments notified of the decisions reached. Following approval by the Y14 Standards Committee and the Co-Secretariats, this issue was approved by the American National Standards Institute on December 12, 1973.

Suggestions for improvement of this standard will be welcome. They should be sent to the American National Standards Institute, 1430 Broadway, New York, N.Y. 10018.

AMERICAN NATIONAL STANDARDS COMMITTEE Y14
Engineering Drawing and Related Documentation Practices

(The following is the Roster of the Committee at the time of approval of this Standard)

OFFICERS

F. L. Spalding, *Chairman*

R. F. Franciose, *Vice-Chairman*   **Leonel V. Porter,** *Secretary*

SECTIONAL COMMITTEE

AMERICAN INSTITUTE FOR DESIGN AND DRAFTING
Francis A. Saint, Tulsa, Oklahoma

AMERICAN SOCIETY OF CIVIL ENGINEERS
F. J. Kircher, Seelye, Stevenson, Value & Knecht, New York, New York

AMERICAN SOCIETY FOR ENGINEERING EDUCATION, THE
R. W. Bokenkamp, University of Illinois, Urbana, Illinois
Kenneth E. Botkin, Purdue University, West Lafayette, Indiana
W. J. Luzadder, Purdue University, West Lafayette, Indiana
H. C. Spencer, Waco, Texas
C. H. Springer, N. Fort Myers, Florida

AMERICAN SOCIETY OF HEATING, REFRIGERATING & AIR CONDITIONING ENGINEERS
H. J. Donovan, alternate, Carrier Corporation, Syracuse, New York
F. Honerkamp, Anemostat Corporation of America, Scranton, Pennsylvania
N. A. LaCourte, alternate, ASHRE, New York, New York

AMERICAN SOCIETY OF MECHANICAL ENGINEERS
A. R. Machell, Xerox Corp., Rochester, New York
F. L. Spalding, University of Illinois, Urbana, Illinois

ASSOCIATION OF AMERICAN RAILROADS
M. F. McCorcle, Springfield, Missouri

BUSINESS EQUIPMENT MFGRS. ASSOCIATION
L. E. Sabine, alternate, Univac Division, Sperry Rand Corp., Philadelphia, Pennsylvania
W. M. Souza, IBM Corp., San Jose, California

DEPT. OF THE ARMY
C. A. Nazian, U.S. Army Ordnance, Philadelphia, Pennsylvania

DEPT. OF COMMERCE–PATENT OFFICE
D. M. Mills, The Commissioner of Patents, Washington, D.C.

DEPT. OF THE NAVY
L. A. Meadows, Naval Ship Engrg. Ctr., Hyattsville, Maryland

ILLUMINATING ENGINEERING SOCIETY
L. E. Barbrow, National Bureau of Standards, Washington, D.C.
J. E. Kaufman, alternate, Illuminating Engineering Society, New York, New York

INSTITUTE OF ELECTRICAL & ELECTRONICS ENGINEERS
Charles A. Fricke, alternate, Philco Ford Division, Willow Grove, Pennsylvania
C. R. Muller, ITT Federal Laboratories, Clifton, New Jersey
G. H. Pearsall, Consolidated Edison Co. of New York, New York, New York

SUBCOMMITTEE 5
DIMENSIONING AND TOLERANCING

P. A. Nicovich, Chairman, Sandia Laboratories, Albuquerque, New Mexico
L. W. Foster, Vice-Chairman, Honeywell, Inc., Minneapolis, Minnesota
D. Bibeau, Lowell, Massachusetts
D. C. Blewitt, Xerox Corp., Rochester, New York
D. J. Buchman, General Electric Co.-Flight Propulsion Div., Cincinnati, Ohio
G. A. Eisner, Western Electric Co., New York, New York
*L. W. Falkner, Warren, Michigan
P. W. E. Gehris, The Pennsylvania State University, Wyomissing, Pennsylvania
J. W. Geier, RCA Corporation, Cherry Hill, New Jersey
H. E. Guetzlaff, John Deere Waterloo Tractor Works, Waterloo, Iowa
J. Hobko, U.S. Army Electronics Command, Fort Monmouth, New Jersey
H. P. Hurd, Naval Air Engrg. Ctr., Philadelphia, Pennsylvania
R. R. Hydell, Detroit Diesel Allison Division, Indianapolis, Indiana
S. J. Levy, Lockheed Missiles & Space Co., Sunnyvale, California
C. A. Nazian, Frankford Arsenal, Philadelphia, Pennsylvania
C. E. Sloan, Detroit Diesel Engine Div., General Motors Corp., Detroit, Michigan
F. L. Spalding, University of Illinois, Urbana, Illinois
W. L. Wein, General Electric Co., Burlington, Vermont
J. L. Zeno, Hq. Air Force Logistics Command, Wright-Patterson AFB, Ohio
J. E. Long, alternate, General Motors Technical Center, Warren, Michigan
C. W. Stockwell, alternate, International Harvester Co., Hinsdale, Illinois

LIAISON REPRESENTATIVES

W. E. Allen, International Standardization Section, Dept. of the Navy, Washington, D.C.
C. R. Austin, Rolls Royce Ltd., Derby, England
V. B. Boulton, Standards Association of Australia, North Sydney, N.S.W., Australia
L. B. Cooper, Ministry of Defence, Ensleigh, Bath, England
R. F. Franciose, General Electric Company, Schenectady, New York
R. Hill, St. Catherines, Ontario, Canada
A. W. Huddleston, Dept. of National Defence, Ottawa, Canada
A. E. Johnson, Moore Special Tool Co., Inc., Bridgeport, Connecticut

---

*Deceased

CONTENTS

AMERICAN NATIONAL STANDARD

ENGINEERING DRAWING AND RELATED DOCUMENTATION PRACTICES

Dimensioning and Tolerancing

5-1 GENERAL DIMENSIONING

5-1.1 GENERAL. This subsection establishes rules, principles, and methods of dimensioning and tolerancing used to define the required condition of a part or component on an engineering drawing. Herein are established uniform practices for stating and interpreting these requirements as they relate to U.S. customary (inch) units.

5-1.1.1 Excluded. Architectural and civil engineering drawing practices and graphical symbols are excluded.

5-1.2 FIGURES. The figures given in this standard illustrate meanings only. In some instances figures have been over-detailed for emphasis, and in other instances figures are by intent incomplete. Numerical values of dimensions and tolerances are illustrative only. If inconsistencies between the text and illustrations are noted, the text takes precedence.

5-1.3 NOTES. Notes in capital letters are intended to appear on finished drawings. Notes in lower case letters are explanatory only and not intended to appear on drawings.

5-1.4 REFERENCE TO THIS STANDARD. When drawings are based on this standard, this fact shall be noted on the drawings or in a document referenced on the drawings.

5-1.5 REFERENCE TO GAGING. This document is not intended as a gaging standard, and any reference to gaging is for explanatory reasons only.

5-1.6 DEFINITIONS. The following terms are defined as their use applies in this standard.

5-1.6.1 Dimension. A numerical value expressed in appropriate units of measure and indicated on a drawing along with lines, symbols, and notes to define the geometrical characteristic of an object.

5-1.6.2 Basic Dimension. A numerical value used to describe the theoretically exact size, shape or location of a feature or datum target. It is the basis from which

permissible variations are established by tolerances on other dimensions, in notes or by feature control symbols.

5-1.6.3 True Position. The theoretically exact location of a feature established by basic dimensions.

NOTE: The former term "true position tolerancing" has been replaced by "positional tolerancing" in this standard.

5-1.6.4 Reference Dimension. A dimension usually without tolerance, used for information purposes only. It does not govern production or inspection operations. A reference dimension is derived from other values shown on the drawing or on related drawings.

5-1.6.5 Datum. Points, lines, planes or cylinders, and other geometric shapes assumed to be exact for purposes of computation, from which the location or geometric relationship (form) of features of a part may be established.

5-1.6.6 Datum Target. A specified point, line, or area on a part used to establish a datum.

5-1.6.7 Feature. A feature is any component portion of a part that can be used as basis for a datum. An individual feature may be:

(a) A plane surface (in which case there is no consideration of feature size).

(b) A single cylindrical or spherical surface, or two plane parallel surfaces (all of which are associated with a size dimension).

Complex features are composed of two or more individual features as defined above.

5-1.6.8 Nominal Size. The designation used for the purpose of general identification.

5-1.6.9 Actual Size. The measured size.

1

5-1.6.10 Limits of Size. The applicable maximum and minimum sizes.

5-1.6.11 Maximum Material Condition (MMC). The condition where a feature of size contains the maximum amount of material within the stated limits of size. For example, minimum hole diameter, maximum shaft diameter.

5-1.6.12 Least Material Condition (LMC). The condition where a feature of size contains the least amount of material within the stated limits of size. For example, maximum hole diameter, minimum shaft diameter.

5-1.6.13 Regardless of Feature Size (RFS). The term used to indicate that a form or positional tolerance applies as follows:

(a) At any increment of size of the feature within its size tolerance.

(b) At the actual size of a datum feature.

5-1.6.14 Virtual Condition. The boundary generated by the collective effects of the MMC limit of a feature and any applicable form or positional tolerance.

5-1.6.15 Allowance. The intentional difference between the MMC limits of size of mating parts. It is the minimum clearance (positive allowance) or maximum interference (negative allowance) between such parts.

5-1.6.16 Tolerance. The total amount by which a specific dimension is permitted to vary. The tolerance is the difference between the maximum and minimum limits.

5-1.6.17 Unilateral Tolerance. A tolerance in which variation is permitted in one direction from the specified dimension.

5-1.6.18 Bilateral Tolerance. A tolerance in which variation is permitted in both directions from the specified dimension.

5-1.6.19 Fit. The general term used to signify range of tightness or looseness which results from application of a specific combination of allowances and tolerances in mating parts.

5-1.6.20 Full Indicator Movement (FIM). The total movement of the indicator when applied to a surface in an appropriate manner. The terms Full Indicator Reading (FIR) and Total Indicator Reading (TIR) were formerly used in previous editions of this standard and had identical meaning to FIM.

5-1.7 FUNDAMENTAL RULES FOR DIMENSIONING. Dimensioning of the geometric characteristics of parts shall clearly define the engineering intent and shall conform to the following rules:

(a) Each dimension shall have a tolerance, either applied directly or indicated by a general note. See American National Standard for Drawing Sheet Size and Format, Y14.1-1957. Those specifically identified as reference, basic, maximum, or minimum dimensions are exceptions to this rule.

(b) Dimensions for size, form and location of features shall be complete to the extent there is full understanding of the characteristics of each feature. Neither scaling (measuring the size of a feature directly from an engineering drawing) nor assumption of a distance or size is permitted.

NOTE: Undimensioned drawings (for example: loft, printed wiring, templates, master layouts, tooling layout) prepared on stable material are excluded, provided the necessary control dimensions are specified.

(c) Dimensions shall be shown between points, lines, or surfaces having a necessary and specific relationship to each other or controlling the location of other components or mating parts.

(d) Dimensions shall be selected and arranged to avoid accumulation of tolerances and more than one interpretation.

(e) The drawing should define a part without specifying manufacturing methods. Thus, only the diameter of a hole is given without indication as to whether it may be drilled, reamed, punched, or made by any other operation. However, in those instances where manufacturing, processing, quality assurance, or environmental information is essential to the definition of engineering requirements, it shall be specified on the drawing or in a document referenced on the drawing.

(f) It is permissible to identify as nonmandatory certain processing dimensions that provide for finish allowance, shrink allowance, and other requirements, provided the final dimensions are given on the drawing or on a higher assembly drawing. Nonmandatory processing dimensions shall be identified by an appropriate note, such as NONMANDATORY (MFG DATA).

(g) Dimensions shall be selected to provide required information. Dimensions shall be shown in true profile views and refer to visible outlines.

(h) Wires, cables, sheets, rods and other materials manufactured to gage or code numbers shall be specified by linear dimensions indicating the diameter or thickness. Gage or code numbers may be shown in parentheses following the dimension.

(i) Surfaces or centerlines shown on drawings at right angles to each other are implied to be 90° apart, without specifying the 90° on the drawing.

5-1.8 UNITS OF MEASUREMENT. The unit of measurement selected should be that unit most compatible with the user of the drawing.

5-1.8.1 US Customary Linear Units. The commonly used US (United States) customary linear unit used on engineering drawings is the inch.

5-1.8.2 SI (Metric) Linear Units. The commonly used SI (International System of Units) linear unit used on engineering drawings is the millimetre.

5-1.8.3 Identification of Linear Units. On drawings where all dimensions are either in inches or millimetres, individual identification of units is not required. However, the drawing shall contain a note stating UNLESS OTHERWISE SPECIFIED, ALL DIMENSIONS ARE IN INCHES (or millimetres as applicable).

5-1.8.3.1 Where some millimetres are shown on an inch-dimensioned drawing, the millimetre value should be followed by the symbol mm. Where some inches are shown on a millimetre-dimensioned drawing, the inch value should be followed by the abbreviation IN.

5-1.8.4 Dual Linear Units. On drawings where linear units in both inches and millimetres are shown as adjacent values for dimensions, the practice specified in subsection 5-7, Dual Dimensioning, is recommended.

5-1.8.4.1 As an alternate, dual linear units may be expressed in tabular form. In such cases, the drawing denotes and is dimensioned in the system of units selected for the design. Dimensional values are duplicated in a table provided on the drawing, where converted values in other units are given. Appropriate captions identify the columns for millimetre and inch entries.

5-1.8.5 Angular Units. Angular dimensions are expressed in degrees, minutes and seconds. These are expressed by symbols; for degrees °, for minutes ′, and for seconds ″. Where degrees are indicated alone, the numerical value shall be followed by the symbol °. Where only minutes or seconds are specified, the number of minutes or seconds shall be preceded by 0° or 0°0′, as applicable. Where desired, the angle may be given in degrees and decimal parts of a degree and the tolerance in decimal parts of a degree. See Figure 1.

5-1.9 TYPES OF DIMENSIONING. Dimensions may be expressed as decimals or common fractions. Decimal dimensioning is preferred. Combination dimensioning employs decimals for all dimensions except the designation of nominal sizes of parts or features such as bolts, screw threads, keyseats, or other standardized fractional designations.

5-1.9.1 Decimal Inch Dimensioning. Complete decimal dimensioning employs decimals for all dimensions and designations. See American National Standard for Decimal Inch, B87.1-1965.

5-1.9.1.1 When specifying decimal inch dimensions on drawings, the following rules shall be observed:

(a) For values less than one, zeros are not used before the decimal point.

(b) In plus and minus tolerancing, the specified dimension contains the same number of decimal places as the tolerance applicable to the dimension. See Figure 2. This applies whether the tolerance is specified directly to the dimension, or in a tolerance note.

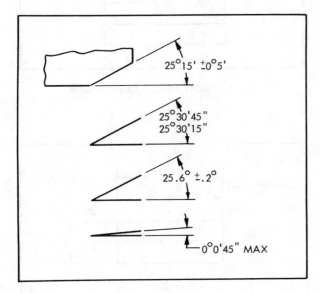

FIG. 1 ANGULAR UNITS

(c) For limit dimensioning, the high and low limits contain the same number of decimal places.

(d) Tolerances are not assumed from the number of decimal places in the dimension.

5-1.9.1.2 Where a decimal value is to be rounded off to a lesser number of places than the total number available, the procedure in American National Standard, Rules for Rounding Off Numerical Values, Z25.1-1940 (R1961), is recommended.

5-1.9.1.3 Decimal points must be uniform, dense and large enough to be clearly visible and meet the reproduction requirements of American National Standard for Line Conventions and Lettering, Y14.2-1973. Decimal points are placed in line with the bottom of the associated digits.

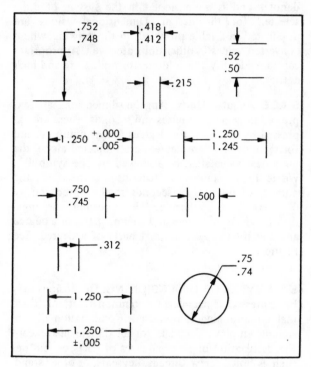

FIG. 2   DECIMAL INCH DIMENSIONS

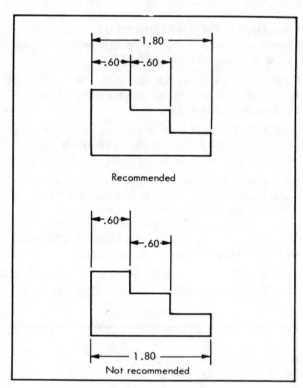

Recommended

Not recommended

FIG. 4   GROUPING OF DIMENSIONS

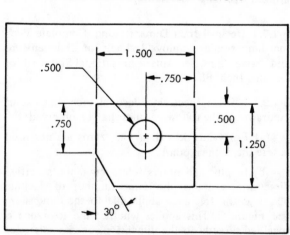

FIG. 3   PLACEMENT OF DIMENSIONS

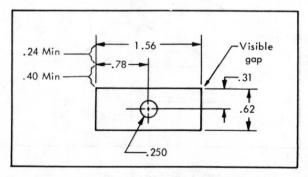

FIG. 5   SPACING OF DIMENSIONS

5-1.10 APPLICATION OF DIMENSIONS. Dimensions are applied by means of dimension lines, extension lines, and leader lines from a dimension, note or specification directed to the appropriate features. See Figures 3 and 5. General notes are used to convey additional information.

5-1.10.1 Dimension Lines. A dimension line, with its arrowheads, shows the direction and extent of a dimension. Numerals indicate the number of units of a measurement. Dimension lines should be broken for insertion of numerals as shown in Figures 2 and 3.

5-1.10.1.1 Dimension lines shall be aligned if practicable and grouped for uniform appearance. See Figure 4.

5-1.10.1.2 Dimension lines are drawn parallel to the direction of measurement. The space between the first dimension line and the part outline should not be less than 0.40 inch; the space between succeeding parallel dimension lines should not be less than 0.24 inch. See Figure 5.

NOTE: These spacings are intended as guides only. If the drawing meets the reproduction requirements of American National Standard Y14.2-1973, nonconformance to these spacing requirements is not basis for drawing rejection.

Where there are several parallel dimension lines, the numerals should be staggered for easier reading. See Figure 6.

5-1.10.1.3 The following shall not be used as a dimension line: a centerline, an extension line, a line that is part of the outline of the object, or a continuation of any of these lines. See Figure 7. A dimension line is not used as an extension line. Exceptions to the above are applicable to the coordinate dimensioning of irregular outlines only. See Figures 38 and 39.

5-1.10.1.4 The dimension line of an angle is an arc drawn with its center at the apex of the angle and terminating at the extensions of the two sides. See Figure 8.

5-1.10.1.5 Crossing dimension lines should be avoided. Where unavoidable, the dimension lines are unbroken.

5-1.10.2 Extension Lines. Extension lines are used to indicate the extension of a surface or point to a location outside the part outline. Normally, extension lines start with a short visible gap from the outline of the part and extend beyond the outermost related dimension line. See Figure 5. Extension lines are usually drawn perpendicular to dimension lines. Where space is limited, extension lines may be drawn at an oblique angle to clearly illustrate where they apply. Where oblique lines are used, the dimension lines are shown in the direction in which they apply. See Figure 9.

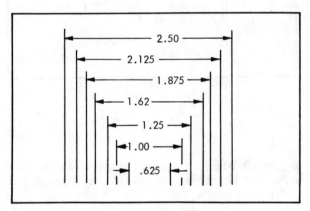

FIG. 6 STAGGERED DIMENSIONS

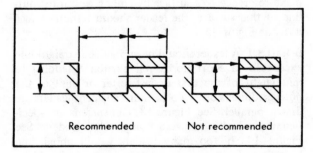

FIG. 7 LOCATION OF DIMENSIONS

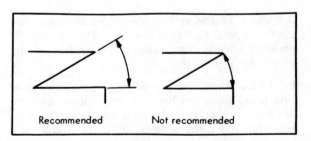

FIG. 8 LOCATION OF DIMENSIONS

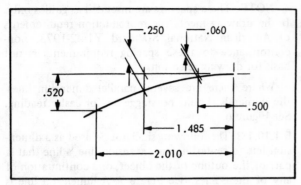

FIG. 9   OBLIQUE EXTENSION LINES

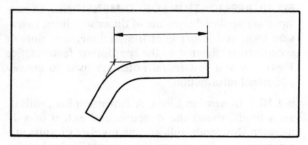

FIG. 12   POINT LOCATIONS

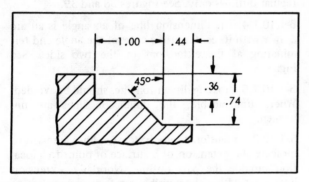

FIG. 10   CROSSING EXTENSION LINES

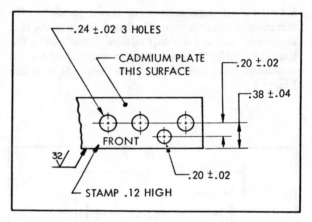

FIG. 13   LEADERS

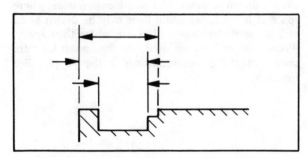

FIG. 11   BREAKS IN EXTENSION LINES

5-1.10.2.1 Extension lines should neither cross one another nor cross dimension lines. The shortest dimension line is shown nearest the outline of the object. See Figure 10.

5-1.10.2.2 Where extension lines cross other extension lines, dimension lines, or object lines, they are not broken. See Figure 10. In instances where extension lines do cross arrowheads or dimension lines close to arrowheads, a break in the extension line is permitted. See Figure 11.

5-1.10.2.3 Where a point is located by extension lines only, the extension lines should pass through the point. See Figure 12.

5-1.10.3 **Leader Line.** A leader line (leader) is used to direct a dimension, note, or symbol to the intended place on the drawing. Normally, a leader terminates in an arrowhead. However, where it is intended a leader make reference to a surface by ending within the outline of that surface, the leader should terminate in a dot. See Figure 13.

5-1.10.3.1 A leader should be an inclined straight line except for a short horizontal portion extending to mid-height of the first or last letter or digit of the note or dimension. Two or more adjacent leaders are drawn parallel. See Figure 13. Dimensions are specified individually to avoid complicated leaders. See Figure 14. If too many leaders would impair the legibility of the drawing, letter symbols should be used to identify features. See Figure 15.

5-1.10.3.2 Where a leader is directed to a circle or arc, its direction should be radial. See Figure 16.

5-1.10.3.3 Avoid the following:

(a) Crossing leaders

(b) Long leaders.

(c) Leaders in a horizontal or vertical direction.

(d) Leaders parallel to adjacent dimension lines, extension lines, or section lines.

(e) Small angles between leaders and the lines where they terminate. See Figure 17.

5-1.10.4 Reading Direction. Dimensions on drawings are either unidirectional or aligned. Unidirectional is preferred. See Figure 18.

5-1.10.4.1 Unidirectional dimensions are placed to be read from the bottom of the drawing.

5-1.10.4.2 Aligned dimensions are placed parallel to their dimension lines. Numerals are read from the bottom or right side of the drawing.

5-1.10.4.3 In both methods, dimensions and notes shown with leaders are aligned with the bottom of the drawing.

5-1.10.5 Reference Dimensions. The preferred method for indicating reference dimensions on drawings is to enclose the dimensions with parentheses. See Figures 19 and 20. As an alternate, reference dimensions may be indicated by the abbreviation REF placed directly following or under the dimension.

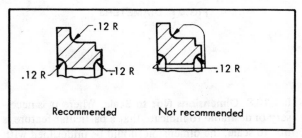

FIG. 14 LEADERS

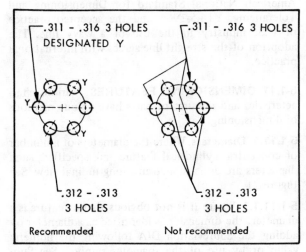

FIG. 15 LEADERS

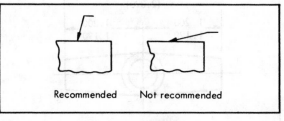

FIG. 17 LEADERS

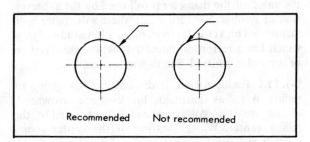

FIG. 16 LEADERS

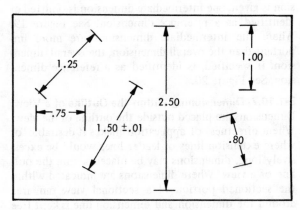

FIG. 18 UNIDIRECTIONAL DIMENSIONS

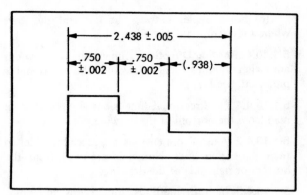

FIG. 19 REFERENCE DIMENSIONS

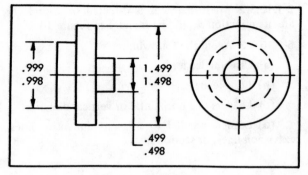

FIG. 21 DIAMETERS

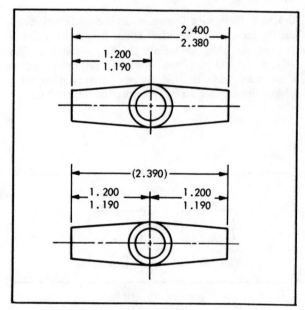

FIG. 20 OVERALL DIMENSIONS

5-1.10.6 Overall Dimensions. Where an overall dimension is given, one intermediate dimension is omitted or identified as a reference dimension. See Figure 19. Where the intermediate dimensions are more important than the overall dimension, the overall dimension, if specified, is identified as a reference dimension. See Figure 20.

5-1.10.7 Dimensioning Within the Outline of a View. Dimensions are placed outside the outline of the view. Where directness of application makes it desirable, or where extension lines or leader lines would be excessively long, dimensions may be placed within the outline of a view. Where dimensions are indicated within the sectioned portion of a sectional view, an area around the dimension and dimension line is kept free from section lines.

5-1.10.8 Dimensions Not to Scale. Where it is necessary or desirable to indicate that a particular feature is not to scale, the dimension should be underlined with a straight line.

NOTE: This is a change from previous editions of American National Standard for Dimensioning and Tolerancing, Y14.5-1966 and the accepted practice of U.S. industry in the use of a wavy line. The adoption of the straight line agrees with international practice.

5-1.11 DIMENSIONING FEATURES. Various characteristics and features of parts have unique methods of dimensioning.

5-1.11.1 Diameters. Where the diameters of a number of concentric cylindrical features are specified, such diameters are dimensioned in a longitudinal view. See Figure 21.

5-1.11.1.1 Where it is not obvious that a feature is a diameter, the dimension is identified as a diameter by adding the abbreviation DIA following the numeric value or by use of the diameter symbol. See Subsection 5-3, Symbology. On a single view drawing on which the circularity of the feature is not illustrated, the value of the diameter is followed by the abbreviation or symbol. See Figure 22. Where a diameter is dimensioned on a circular view, or on a longitudinal view which has a reciprocal circular view, the abbreviation or symbol is omitted. See Figures 21 and 23.

5-1.11.2 Radii. An arc is dimensioned by giving its radius. A radius dimension line uses one arrowhead, at the arc end. An arrowhead is never used at the radius center. Where location of the center is important and space permits, a dimension line is drawn from the radius center with the arrowhead touching

8

the arc, and the dimension placed between the arrowhead and the center. Each radius numeric value should be followed by the abbreviation R. Where space is limited, the radial dimension line is extended through the radius center. Where it is inconvenient to place the arrowhead between the radius center and the arc, it may be placed outside the arc with a leader. See Figure 24.

5-1.11.2.1 Where a dimension is given to the center of a radius, a small cross should be drawn at the center. Extension lines and dimension lines are used to locate the center. See Figure 25. Where the location of the center is unimportant, an arc is located by tangent lines with the radius center not shown. See Figure 26.

5-1.11.2.1.1 Where the center of a radius is outside the drawing or interferes with another view, the radius dimension line may be shortened. See Figure 27. The portion of the dimension line next to the arrowhead is radial relative to the curved line. Where the radius dimension line is shortened and the center located by coordinate dimensions, the dimension line locating the center should be shortened.

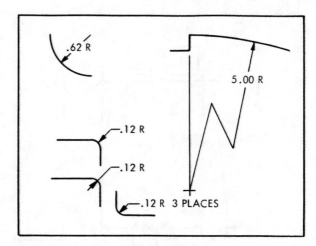

FIG. 24 RADII

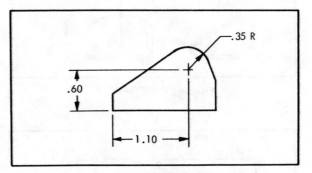

FIG. 25 LOCATED RADIUS CENTER

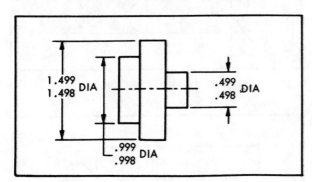

FIG. 22 DIAMETERS

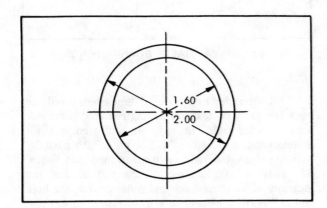

FIG. 23 DIAMETERS

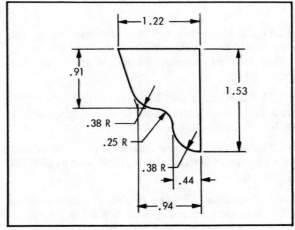

FIG. 26 RADII LOCATED BY TANGENTS

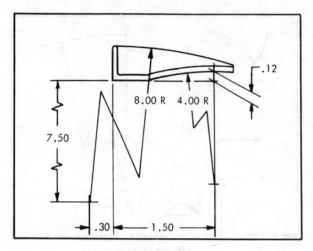

FIG. 27 SHORTENED RADII

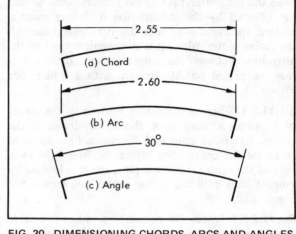

FIG. 30 DIMENSIONING CHORDS, ARCS AND ANGLES

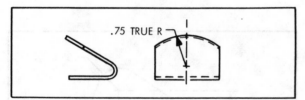

FIG. 28 TRUE RADIUS

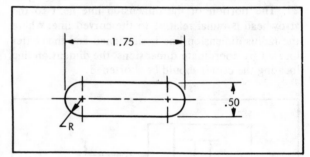

FIG. 31 ROUNDED ENDS

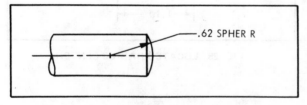

FIG. 29 SPHERICAL RADIUS

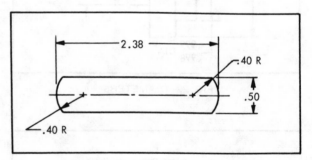

FIG. 32 PARTIALLY ROUNDED ENDS

5-1.11.2.2 Where the radius is dimensioned in the view that does not show the true shape of the radius, TRUE R is added after the radius dimension. See Figure 28.

5-1.11.2.3 Where a part has a number of radii of the same dimension, a note is used in lieu of dimensioning each radius separately.

5-1.11.2.4 Where spherical surfaces are dimensioned by radii, the radius dimension is followed by SPHER R. See Figure 29.

5-1.11.3 Chords, Arcs and Angles. The dimensioning of chords, arcs, and angles should be as shown in Figure 30. Where required for clarity, the dimension shall be modified with a term such as ARC or CHORD.

5-1.11.4 Rounded Ends. Overall dimensions should be used for parts or features having rounded ends. For fully rounded ends, the radii are indicated, but not dimensioned. See Figure 31. For parts with partially rounded ends, the radii are dimensioned. See Figure 32. Where a hole location is more critical than the location of a radius from the same center, the hole and radius are dimensioned and toleranced separately. See Figure 33.

10

5-1.11.5 Rounded Corners. Where corners are rounded, dimensions locate the edges, and arcs are tangent. See Figure 34.

5-1.11.6 Outlines Consisting of Arcs. A curved outline composed of two or more arcs is dimensioned by giving the radii of all arcs and locating the necessary centers with coordinate dimensions, other radii being located on the basis of their points of tangency. See Figure 35.

5-1.11.7 Irregular Outlines. Irregular outlines may be dimensioned by several methods. See Figures 36 through 39.

5-1.11.7.1 Circular or noncircular outlines are dimensioned by the rectangular coordinate or offset method. See Figure 36. Extend each dimension line to the datum line. Where many coordinates are required to describe a contour, the vertical and horizontal coordinate dimensions are tabulated. See Figure 37.

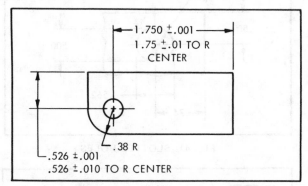

FIG. 33 MULTIPLE TOLERANCES

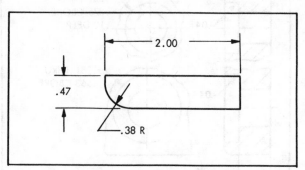

FIG. 34 ROUNDED CORNERS

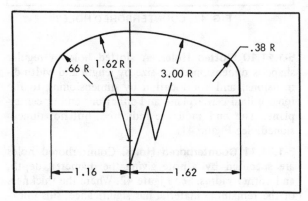

FIG. 35 ARCS

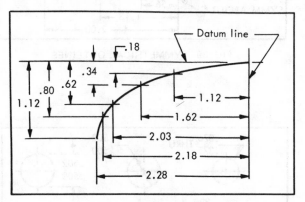

FIG. 36 IRREGULAR OUTLINES

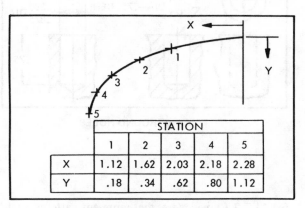

STATION					
	1	2	3	4	5
X	1.12	1.62	2.03	2.18	2.28
Y	.18	.34	.62	.80	1.12

FIG. 37 TABULATED OUTLINES

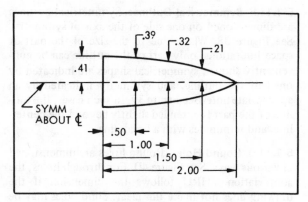

FIG. 38 SYMMETRICAL OUTLINES

11

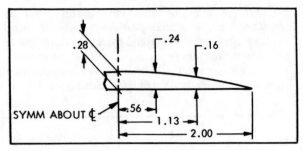

FIG. 39   SYMMETRICAL OUTLINES

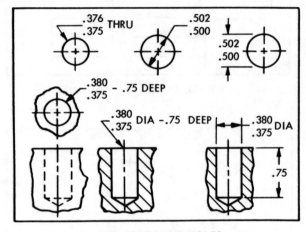

FIG. 40   ROUND HOLES

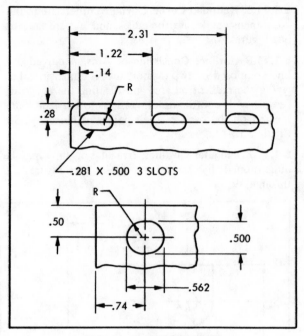

FIG. 41   SLOTTED HOLES

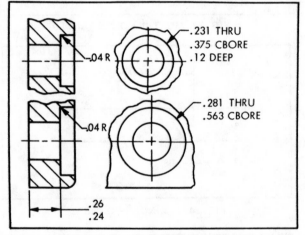

FIG. 42   COUNTERBORED HOLES

5-1.11.7.2 Curved pieces that represent patterns, such as, leather holsters, vehicle covers, etc., are delineated by a grid system with numbered datum lines.

5-1.11.8 Symmetrical Outlines. Symmetrical outlines are dimensioned on one side of the axis of symmetry. See Figure 38. Where due to the size of the part or space limitations, only part of the outline can be conveniently shown, symmetrical shapes are indicated by one-half the outline, and symmetry indicated by an appropriate note. See Figure 39. In such cases, the outline of the part is extended slightly beyond the center-line, and terminates with a break-line.

5-1.11.9 Round Holes. Round holes are dimensioned in various ways. See Figure 40. For through holes, the abbreviation THRU follows the dimension if the drawing does not make this clear. Blind holes may be dimensioned as illustrated in Figure 40.

5-1.11.10 Slotted Holes. A slotted hole of regular shape is dimensioned for size by length and width dimensions, and for location by dimensioning to the longitudinal centerplane and either one end or center-plane. The end radii are indicated, but not dimensioned. See Figure 41.

5-1.11.11 Counterbored Holes. Counterbored holes are specified by a note giving the diameter, depth, and corner radius. See Figure 42. Where the thickness of the remaining material has significance, this thickness rather than the depth is dimensioned.

5-1.11.12 Countersunk and Counterdrilled Holes. For countersunk holes, the diameter and included angle of the countersink are specified. For counterdrilled holes, the diameter, depth, and included angle of the counterdrill are specified. See Figure 43. The depth dimension is the depth of the full diameter of the counterdrill.

5-1.11.12.1 Where a circular chamfer or countersink intersects a curved surface, the dimension specified on the drawing applies at the minor diameter of the chamfer or countersink. See Figure 44.

5-1.11.13 Spotfaces. For spot-faced holes, the diameter of the flat area and either the depth or the remaining thickness of material should be specified. See Figure 45. A spotface may be specified by note only, and need not be delineated on the drawing. If no depth or remaining thickness of material is specified, the spot-facing is the minimum depth necessary to clean up the surface to the specified diameter.

5-1.11.14 Machining Centers. Where machining centers are to remain on the finished part, they are indicated or dimensioned on the drawing.

5-1.11.15 Chamfers. Chamfers are dimensioned by an angle and a length. See Figure 46. The linear dimension is the measurement along the length of the part.

5-1.11.15.1 A note may be used to specify 45° chamfers, as in Figure 47. This method is used only with 45° chamfers as the linear value applies to either the longitudinal or radial dimension.

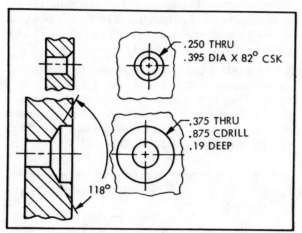

FIG. 43 COUNTERSUNK AND COUNTERDRILLED HOLES

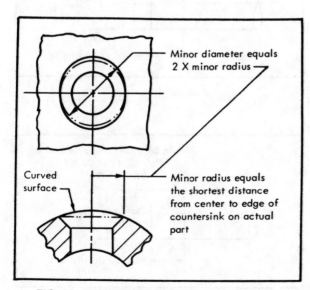

FIG. 44 CURVED SURFACE INTERSECTED BY COUNTERSINK

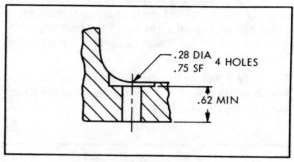

FIG. 45 SPOT-FACED HOLES

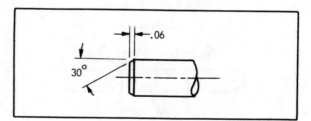

FIG. 46 CHAMFERS

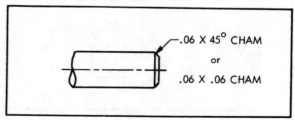

FIG. 47 45 DEGREE CHAMFER

5-1.11.15.2 Where the edge of a round hole is chamfered, the practice of 5-1.11.15.1 is followed except where the chamfer diameter requires dimensional control. See Figure 48.

5-1.11.16 Keyseats. Keyseats are dimensioned by width, depth, location, and if required, length. The depth is dimensioned from the opposite side of the shaft or hole. See Figure 49.

5-1.11.17 Knurling. Knurling is specified in terms of type, pitch, and diameter before and after knurling. Where control is not required, the diameter after knurling is omitted. Where only portions of a feature require knurling, axial dimensioning must be provided. See Figure 50.

5-1.11.17.1 Where required to provide a press-fit between parts, knurling is specified by a note on the drawing which includes the type of knurl required, its pitch, the toleranced diameter of the feature prior to knurling and the minimum acceptable diameter after knurling. See Figure 51.

5-1.11.17.2 For additional information on specifying knurling, see American National Standard for Knurling, B94.6-1966 (R1972).

5-1.11.18 Rods and Tubing Details. Rods and tubing are dimensioned in three coordinate directions or by specifying the straight lengths, bend radii, angles of bend, and angles of twist for all portions of the item. This may be done by means of auxiliary views, by tabulation, or by supplementary data.

5-1.11.19 Screw Threads. Methods of specifying and dimensioning screw threads are covered in American National Standard for Screw Threads, Y14.6-1957.

5-1.11.20 Surface Texture. Methods of specifying surface roughness, waviness and lay are covered in American National Standard for Surface Texture, B46.1-1962 (R1971).

5-1.11.21 Gears. Methods of specifying gear requirements are covered in American National Standards for Gears, Splines and Serrations, Y14.7-1958 and Gear Drawing Standards–Part 1: for Spur, Helical, Double Helical & Rack, Y14.7.1-1971.

5-1.12 LOCATION. Linear or angular dimensions locate features with respect to one another or from a datum. Feature to feature distances may be adequate for describing certain parts. Dimensions from a datum may be necessary if a part with more than one critical dimension must mate with another part.

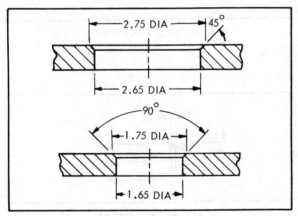

FIG. 48   HOLE CHAMFERS REQUIRING DIAMETER CONTROL

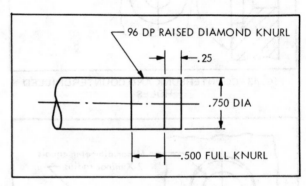

FIG. 50   KNURLS

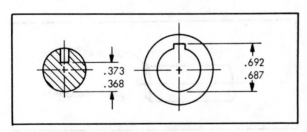

FIG. 49   KEYSEATS

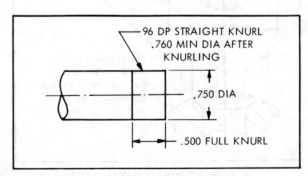

FIG. 51   KNURLS FOR PRESS FITS

5-1.12.1 Locating Symmetrical Features. Figures 52 through 60 illustrate the positioning of round holes or other features of symmetrical contour by giving distances, or distances and directions to the feature centers.

5-1.12.2 Repetitive Dimensions. Where a series of holes or other features are spaced equally, the notation EQUALLY SPACED, EQL SP, or XX SPACES AT .XX may be used in the note pertaining to the features. See Figures 55, 56 and 57. Any such method of dimensioning should include the appropriate specification of applicable tolerances.

5-1.12.3 Rectangular Coordinate Dimensioning. Rectangular coordinate dimensioning is where all dimensions are measured from two or three mutually perpendicular datum planes.

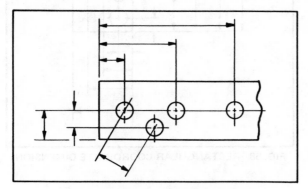

FIG. 52 COORDINATE DIMENSIONS

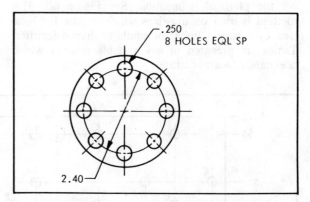

FIG. 55 REPETITIVE DIMENSIONS

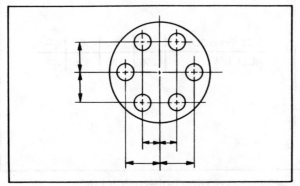

FIG. 53 RECTANGULAR COORDINATED DIMENSIONS

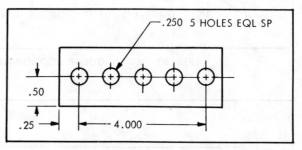

FIG. 56 REPETITIVE DIMENSIONS

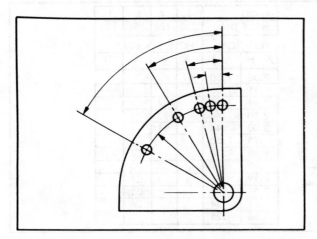

FIG. 54 POLAR COORDINATE DIMENSIONS

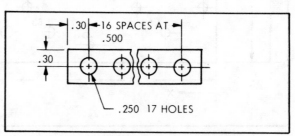

FIG. 57 REPETITIVE DIMENSIONS

15

5-1.12.3.1 Coordinate dimensioning with dimension lines must clearly indicate the features from which the dimensions originate. See Figures 52, 53 and 58.

5-1.12.3.2 Coordinate dimensions without dimension lines originate from datum planes indicated as zero coordinates. Dimensions from these are shown on extension lines without the use of dimension lines or arrowheads. See Figure 59.

5-1.12.4 Tabular Dimensioning. Tabular dimensioning is a type of rectangular coordinate dimensioning in which dimensions from mutually perpendicular datum planes are listed in a table on the drawing rather than on the pictorial delineation. See Figure 60. This method is used on drawings which require the location of a large number of similarly shaped features. Tables are prepared in any suitable manner which adequately locates features.

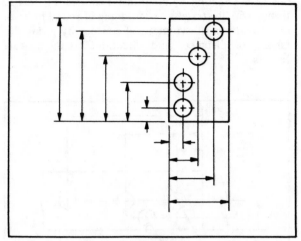

FIG. 58 RECTANGULAR COORDINATE DIMENSIONS

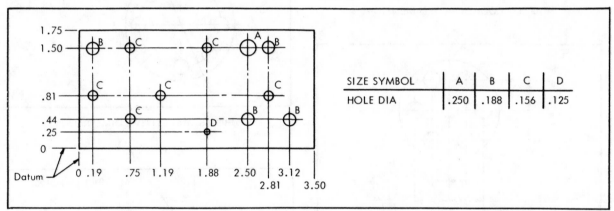

SIZE SYMBOL	A	B	C	D
HOLE DIA	.250	.188	.156	.125

FIG. 59 RECTANGULAR COORDINATE DIMENSIONS WITHOUT DIMENSION LINES

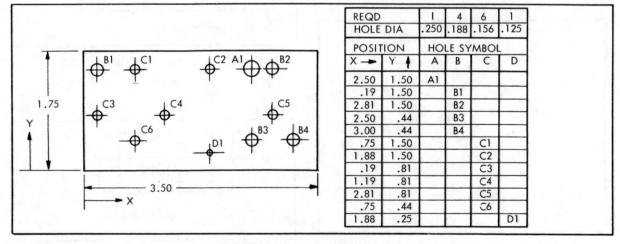

REQD		1	4	6	1
HOLE DIA		.250	.188	.156	.125
POSITION		HOLE SYMBOL			
X →	Y ↑	A	B	C	D
2.50	1.50	A1			
.19	1.50		B1		
2.81	1.50		B2		
2.50	.44		B3		
3.00	.44		B4		
.75	1.50			C1	
1.88	1.50			C2	
.19	.81			C3	
1.19	.81			C4	
2.81	.81			C5	
.75	.44			C6	
1.88	.25				D1

FIG. 60 TABULAR DIMENSIONING

16

5-2 GENERAL TOLERANCING

5-2.1 GENERAL. This subsection establishes practice for indicating tolerances on linear and angular dimensions, applicability for material condition modifiers, and rules for interpreting limits and tolerances.

5-2.1.1 Application. Required tolerances may be expressed as follows:

(a) As specific limits or a tolerance applied directly to the dimension.

(b) In the form of a special note referring to specific dimensions.

(c) As specified in other documents referenced on the drawing for specific features or processes.[1]

(d) In a general tolerance note referring to all dimensions on a drawing for which tolerances are not otherwise specified. See American National Standard Y14.1-1957.

5-2.1.1.1 Tolerances on dimensions that locate features may be applied directly to the locating dimensions or specified by the positional tolerancing method described in Subsection 5-5, Tolerances of Location.

5-2.1.1.2 Tolerances are not applied directly to basic and reference dimensions.

5-2.2 TOLERANCING METHODS. A tolerance applied directly to the dimension may be expressed as follows:

(a) *Limit dimensioning.* The high limit (maximum value) is placed above the low limit (minimum value). See Figure 61. When expressed in a single line the low limit precedes the high limit and both separated by a dash.

(b) *Plus and Minus Tolerancing.* The dimension is given first and is followed by a plus and minus expression of tolerance. See Figure 62.

5-2.3 TOLERANCE EXPRESSION. Tolerance limits or the plus and minus tolerance and its dimension shall be expressed in the same form and with the same number of decimal places. For example:

$$\frac{.750}{.748} \ldots \text{not} \ldots \frac{.75}{.748}$$

$$.500 \pm .005 \ldots \text{not} \ldots .50 \pm .005$$

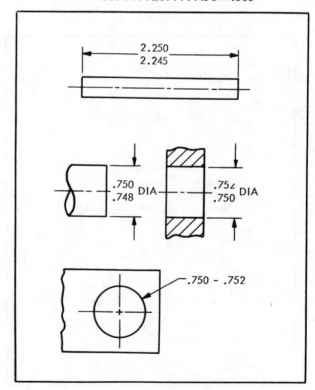

FIG. 61   LIMIT DIMENSIONING

5-2.4 INTERPRETATION OF LIMITS. All limits are absolute. Dimensional limits, regardless of the number of decimal places, are used as if they were continued with zeros. For example:

| 1.22 | Means | 1.220---0 |
| 1.20 | | 1.200---0 |

| 1.202 | Means | 1.2020---0 |
| 1.200 | | 1.2000---0 |

For the purpose of determining conformance within limits, the measured value is compared directly with the specified value and any deviation outside the specified limiting value signifies nonconformance with the limits.

5-2.4.1 Effect of Surface Coatings. Where a part is to be plated or coated, the applicability of dimensions before or after coating shall be specified on the drawing or in a referenced document.[1] This also applies to the applicability of surface texture ratings. Typical examples of notes are the following:

---

[1] Engineering drawings prepared by or for the Department of Defense shall only reference documents approved by the cognizant DoD contracting officer or other approving agency.

(a) Dimensional limits apply after plating.

(b) Dimensional limits apply before plating.

(c) Dimensional limits and surface texture designations apply after plating.

(d) Dimensional limits and surface texture designations apply before plating.

(For other coatings, substitute the appropriate expression for plating.)

5-2.5 SINGLE LIMITS. MIN or MAX is placed after a numeral where other elements of design definitely determine the other unspecified limit. Features such as depths of holes, lengths of threads, corner radii, chamfers, etc., may be limited this way. Single limits are used where (a) the intent will be clear, and (b) the unspecified limit can approach zero or infinity and will not result in conditions detrimental to the design.

5-2.6 TOLERANCE ACCUMULATION. Figure 63 compares the tolerance accumulation resulting from three different methods of dimensioning.

(a) *Chain dimensioning.* The maximum variation between any two features is equal to the sum of the tolerances on the intermediate distances. See Figure 63(a). (This results in the greatest tolerance accumulation as illustrated by the ± .003 variation between holes X and Y.)

(b) *Datum dimensioning.* The maximum variation between any two features is equal to the sum of the tolerances on the two dimensions from the datum to the features. See Figure 63(b). (This reduces the tolerance accumulation as illustrated by the ± .002 variation between holes X and Y.)

(c) *Direct dimensioning.* The maximum variation between any two features is controlled by the tolerance on the dimension between the features. See Figure 63(c). (This results in the least tolerance accumulation as illustrated by the ± .001 variation between holes X and Y.)

5-2.7 ANGULAR SURFACES. An angular surface may be defined by either a combination of a linear dimension and an angle; or by linear dimensions alone. Each arrangement has the effect of specifying a particular tolerance zone within which the surface must lie. Where an angular surface is defined by a

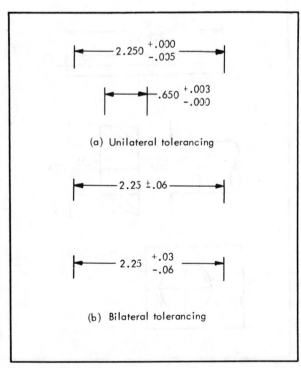

(a) Unilateral tolerancing

(b) Bilateral tolerancing

FIG. 62   PLUS AND MINUS TOLERANCING

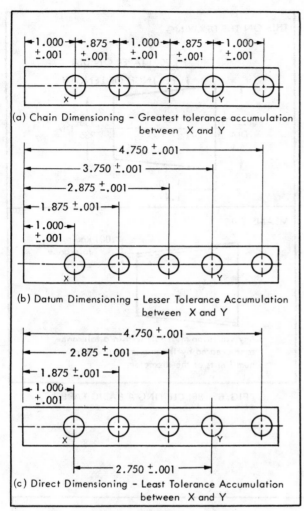

FIG. 63 DIMENSIONING COMPARISON

(a) Chain Dimensioning - Greatest tolerance accumulation between X and Y

(b) Datum Dimensioning - Lesser Tolerance Accumulation between X and Y

(c) Direct Dimensioning - Least Tolerance Accumulation between X and Y

5-2.8 CONICAL TAPERS. A conical taper may be defined by suitable combination of the following dimensions and tolerances. (See Figures 66 through 69):

(a) The diameter at one end of the tapered feature.

(b) The length of the tapered feature.

(c) The diameter at a selected cross-sectional plane which may be within or outside of the tapered feature. The position of this plane is shown with a basic dimension.

(d) A dimension locating a cross-sectional plane at which a basic diameter is specified.

(e) The rate of taper.

(f) The included angle.

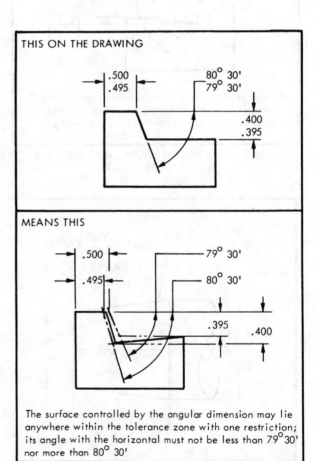

THIS ON THE DRAWING

MEANS THIS

The surface controlled by the angular dimension may lie anywhere within the tolerance zone with one restriction; its angle with the horizontal must not be less than 79° 30' nor more than 80° 30'

FIG. 64 TOLERANCING AN ANGULAR SURFACE USING A COMBINATION OF LINEAR AND ANGULAR DIMENSIONS

combination of a linear and angular dimension, the surface must lie between planes represented by the tolerance zone. See Figure 64. The tolerance zone will be wider as the distance from the apex of the angle increases. In order to avoid a tolerance zone of this shape, the basic angle method is used. In Figure 65, no tolerance is placed on the basic angle. The actual angular variation permitted is defined by the tolerance on the linear dimension plus the basic angle. This defines a tolerance zone with parallel boundaries at the basic angle. The accuracy of the produced angle must be such that no part of the resulting surface exceeds this boundary.

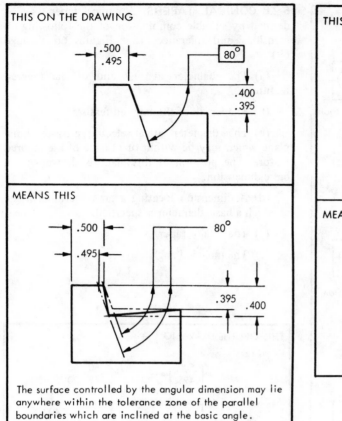

THIS ON THE DRAWING

MEANS THIS

The surface controlled by the angular dimension may lie anywhere within the tolerance zone of the parallel boundaries which are inclined at the basic angle.

FIG. 65 TOLERANCING AN ANGULAR SURFACE
WITH A BASIC ANGLE

THIS ON THE DRAWING

MEANS THIS

Any variation must fall within a tolerance zone created by the maximum and minimum limits of the diameter

FIG. 67 SPECIFYING A BASIC TAPER

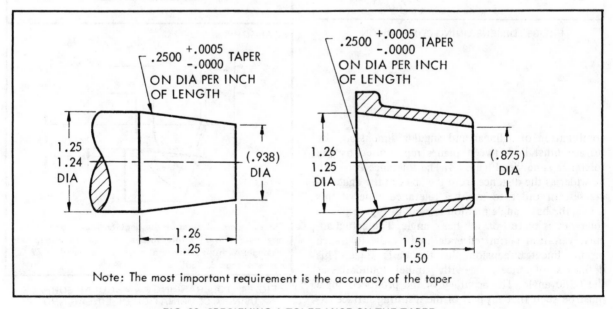

Note: The most important requirement is the accuracy of the taper

FIG. 66 SPECIFYING A TOLERANCE ON THE TAPER

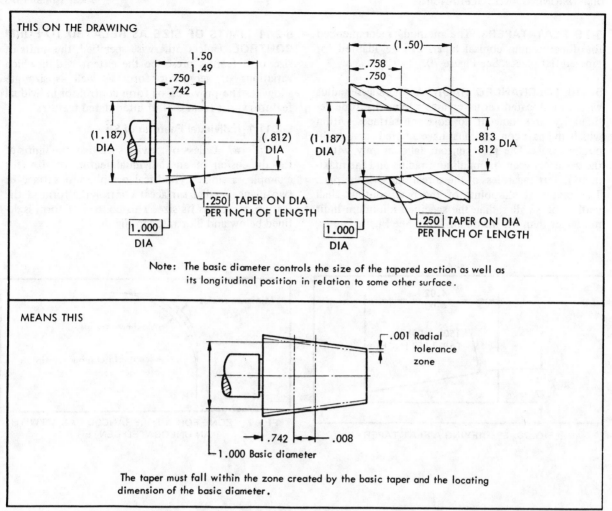

FIG. 68 SPECIFYING A BASIC TAPER AND A BASIC DIAMETER

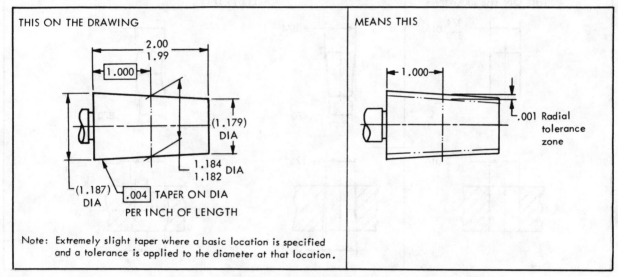

FIG. 69 SPECIFYING A BASIC TAPER AND A BASIC LENGTH

21

5-2.9 FLAT TAPERS. The methods recommended for dimensioning conical tapers can be adapted for tapered flat pieces. See Figure 70.

5-2.10 TOLERANCED RADII. A toleranced radius with an unlocated center creates a tolerance zone defined by arcs tangent to adjacent surfaces within which the part contour must have a faired curve without reversals. The part contour falls entirely within the zone between the minimum radius and the maximum radius, regardless of the actual shape of the part. The radius at all points of the part contour shall neither be smaller than the specified minimum limit nor larger than the maximum limit. See Figure 71.

5-2.11 LIMITS OF SIZE AS RELATED TO FORM CONTROL. Unless otherwise specified, the limits of size of a feature prescribe the extent within which variations of geometric form as well as size are allowed. The provision of form control for individual features varies from that of interrelated features.

5-2.11.1 Individual Features of Size.

Rule #1. Unless otherwise specified, the limits of the dimension of an individual feature of size (for example, a single cylindrical or spherical surface or two plane parallel surfaces) control the form of the feature as well as its size. This control of form is defined below and illustrated in Figure 72:

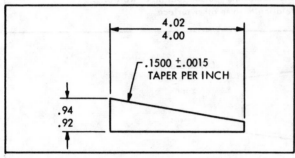

FIG. 70 SPECIFYING A FLAT TAPER

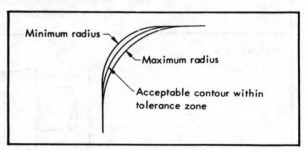

FIG. 71 ZONE FOR A TOLERANCED RADIUS WITH
AN UNLOCATED CENTER

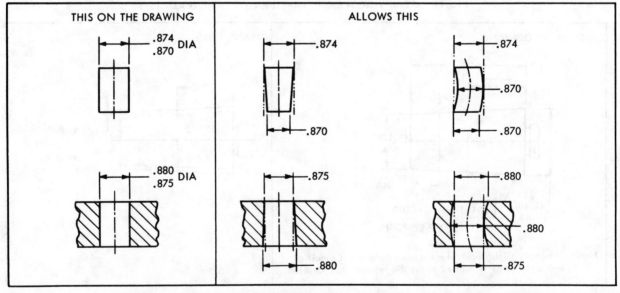

FIG. 72 EXTREME VARIATIONS OF FORM ALLOWED BY A SIZE TOLERANCE

(a) No element of the actual feature shall extend beyond a boundary of perfect form at MMC. This boundary is the true geometric form implied by the drawing.

NOTE: The form control provision of paragraph (a) does not apply to stock such as bars, sheets, and tubing where established standards prescribe straightness, flatness, and other conditions. Where variations of form are not given on drawings of parts made from these materials, standards for the materials govern the surfaces that remain in the "as furnished" condition for the finished part.

(b) The actual size of the feature at any cross-section shall be within the LMC limit of size.

5-2.11.2 Interrelated Features. The form control provision of Rule 1, paragraph(a), applies to individual features and not to the interrelationship of features. Where such control of interrelated features is necessary, one of the following methods is used:

(a) Specify a zero form tolerance at MMC for the features. See 5-6.3.2.

(b) Indicate this control for the features involved by a note such as PERFECT FORM AT MMC REQUIRED FOR INTERRELATED FEATURES.

(c) By relating the dimensions to a datum reference frame.

5-2.11.3 Perfect Form at MMC Not Required. Where it is desired to permit a specified form tolerance to exceed the boundary of perfect form at MMC, a note such as, PERFECT FORM AT MMC NOT REQD, is specified exempting the pertinent size dimension from the provision of Rule 1, paragraph(a).

5-2.12 APPLICABILITY OF MMC OR RFS. Applicability of MMC or RFS is limited to features subject to variations in size. These may be datum features or other features controlled by location and form tolerances. Where a feature or datum feature is one plane surface, MMC or RFS is not applicable. However, MMC or RFS is applicable where the tolerance zone for a feature of size applies to its axis or center plane, not to its surface elements. The following rules apply for indicating MMC or RFS:

Rule #2. For a tolerance of position (formerly called true position tolerance), MMC or RFS is specified on the drawing with respect to the individual tolerance, datum reference, or both, as applicable.

Rule #2a–(Alternate Practice). For a tolerance of position (formerly called true position tolerance), MMC applies with respect to the individual tolerance, datum reference, or both, where no modifier is specified. RFS must be specified on the drawing where it is required.

NOTE: This rule is contrary to international practice and should not be used on drawings subject to international use and interpretation.

Rule #3. For other than a tolerance of position (formerly called true position tolerance), RFS applies with respect to the individual tolerance, datum reference, or both, where no modifier is specified. MMC must be specified on the drawing where it is required.

5-2.12.1 Effect of RFS. Where a location or form tolerance is applied on an RFS basis, the specified tolerance is independent of the size of the considered feature. The tolerance is limited to the specified value regardless of the actual size of the feature. Likewise, referencing a datum feature on an RFS basis means that a centering about its axis or center plane is necessary, irrespective of the actual size of the feature. See 5-4.6.1 and 5-4.6.2.

5-2.12.2 Effect of MMC. Where a tolerance of position or form is applied on an MMC basis, the specified tolerance is interdependent on the size of the considered feature. The tolerance is limited to the specified value if the feature is produced at its MMC limit of size. Where the actual size of the feature has departed from MMC, an increase in the tolerance is allowed equal to the amount of such departure. The total permissible variation in position or form is maximum when the feature is at LMC. Likewise, referencing a datum feature on an MMC basis means the datum is the axis or center plane of a diameter or width fixed at the MMC limit of size of the feature. See 5-4.6.3. Where the actual size of the datum feature has departed from MMC, a deviation is allowed between its axis or center plane and the datum axis or center plane.

5-2.12.3 Effect of Zero Tolerance at MMC. Where a tolerance of position or form is applied on a zero tolerance at MMC basis, the tolerance is totally dependent on the size of the considered feature. No tolerance of position or form is allowed if the feature is produced at its MMC limit of size; and in this case it must be (a) located at true position or (b) be perfect in form, as applicable. Where the actual size of the feature has departed from MMC, a tolerance is allowed equal to the amount of such departure. The total permissible variation in position or form is maximum when the feature is at least material condition, unless a maximum is specified.

5-2.12.3.1 Zero tolerance at MMC is specified where a boundary of perfect form at MMC, as determined by the limits of size, is a requirement of the design. See B6 of Appendix B, and Figures 172 and 173.

5-2.13 SCREW THREADS. The following rule applies to a tolerance of location or form and datum references specified for screw threads:

Rule #4. Each tolerance of form or location and datum reference for a screw thread applies to the pitch diameter.

Where design requirements necessitate an exception to this rule, a qualifying notation (such as MINOR DIA, MAJOR DIA) shall be shown beneath the feature control symbol or beneath the datum identifying symbol, as applicable.

5-2.14 GEARS & SPLINES. Each form or location tolerance and datum reference applied to gears and splines designates the specific feature of the gear or spline to which each applies. See 5-2.13.

5-2.15 VIRTUAL CONDITION. Depending upon its function, a feature is controlled by tolerances, such as size, form, and position, and MMC or RFS may be applied. Consideration must be given to the collective effect of these factors in determining the clearance between mating parts and in establishing gage feature

sizes. From such consideration, a net resultant boundary is derived termed virtual condition, as illustrated by the following:

(a) Where a straightness tolerance of .015 diameter is specified at MMC or RFS for a pin whose size limits are .605-.615 diameter, the virtual condition boundary is .630 diameter.

(b) Where a .010 diameter positional tolerance is specified at MMC for a hole whose size limits are .249-.252, the virtual condition boundary is .239 diameter.

(c) Where a perpendicularity tolerance of .010 diameter is specified at MMC for a pin whose size limits are .500-.504 diameter, the virtual condition boundary is .514 diameter and is perpendicular to the datum.

5-2.15.1 Datum Features at Virtual Condition. A virtual condition exists where a datum feature of size is controlled by a separate tolerance of location or form and the tolerance zone applies to its axis or center plane. The following rule applies to a feature of size that, in addition to serving as a datum feature, has such a tolerance of form or location applied:

Rule #5. Although referenced in a feature control symbol at MMC, a datum feature of size controlled by a separate tolerance of location or form applies at its virtual condition.

Where it is not intended for the virtual condition to apply, an appropriate zero tolerance at MMC should be specified to control the attitude of the datum feature.

5-3 SYMBOLOGY

5-3.1 GENERAL. This subsection establishes the symbols for specifying geometric characteristics on engineering drawings. Symbols should be of sufficient clarity to meet legibility and reproducibility requirements of American National Standard, Y14.2-1973.

5-3.2 USE OF NOTES TO SUPPLEMENT SYMBOLS. Situations may arise in which the precise geometric requirement desired cannot be conveyed by symbols. In such instance, a note is used, either separately or supplementing a symbol, to describe the requirement.

5-3.3 SYMBOL CONSTRUCTION. Information related to the construction, form, and proportion of individual symbols described herein are in Appendix D.

5-3.3.1 Geometric Characteristic Symbols. The symbols denoting geometric characteristics are shown in Figure 73.

5-3.3.2 Datum Identifying Symbol. The datum identifying symbol consists of a frame containing the datum reference letter preceded and followed by a dash. See Figure 74. The symbol is associated with the datum feature by one of the methods prescribed in 5-3.5.

5-3.3.2.1 Letters of the alphabet (except I, O, and Q) are used as datum reference letters. Each datum feature requiring identification should be assigned a different datum reference letter. When datum features requiring identification on a drawing are so numerous as to exhaust the single alpha series, the double alpha series shall be used, that is, AA through AZ.

5-3.3.3 Basic Dimension Symbol. The symbolic means of identifying a basic dimension is to enclose the dimension in a frame, as shown by Figure 75.

5-3.3.4 MMC and RFS Symbols. The symbols used to designate "maximum material condition" and "regardless of feature size" are shown in Figure 76. In notes, the abbreviations MMC and RFS or their spelled-out terms are used.

5-3.3.5 Diameter Symbol. The symbol used to designate a diameter is as shown in Figure 76. It precedes the specified tolerance in a feature control symbol. The symbol may be used elsewhere on a drawing in place of the abbreviation DIA.

5-3.3.6 Projected Tolerance Zone Symbol. The symbol used to designate a projected tolerance zone is as shown in Figure 76.

5-3.3.7 Reference Dimension Symbol. The symbolic means of identifying a reference value is by enclosing each such value with parentheses as shown in Figure 76.

5-3.3.8 Datum Target Symbol. The datum target symbol is a circle divided into four quadrants. The letter placed in the upper left quadrant identifies its associated datum feature. The numeral placed in the lower right quadrant identifies the target. See Figure 77 and 5-4.7.

5-3.4 COMBINED SYMBOLS. Individual symbols, datum reference letters, and the desired tolerance are appropriately combined to express a tolerance symbolically.

5-3.4.1 Feature Control Symbol. A position or form tolerance is stated by means of the feature control symbol consisting of a frame containing the geometric characteristic symbol followed by the allowable tolerance. A vertical line separates the symbol from the tolerance. Where applicable, the tolerance shall be preceded by the diameter symbol and followed by the symbol for MMC or RFS. See Figure 78.

5-3.4.2 Feature Control Symbol Incorporating Datum References. Where a tolerance of position or form is related to a datum, this relationship is stated in the feature control symbol by placing the datum reference letter following either the geometric characteristic symbol or the tolerance. Vertical lines separate these entries. Where applicable, the datum reference letter

		CHARACTERISTIC	SYMBOL	NOTES
INDIVIDUAL FEATURES	FORM TOLERANCES	STRAIGHTNESS	—	1
		FLATNESS	▱	1
		ROUNDNESS (CIRCULARITY)	○	
		CYLINDRICITY	⌭	
INDIVIDUAL OR RELATED FEATURES		PROFILE OF A LINE	⌒	2
		PROFILE OF A SURFACE	⌓	2
RELATED FEATURES		ANGULARITY	∠	
		PERPENDICULARITY (SQUARENESS)	⊥	
		PARALLELISM	//	3
	LOCATION TOLERANCES	POSITION	⊕	
		CONCENTRICITY	◎	3,7
		SYMMETRY	≡	5
	RUNOUT TOLERANCES	CIRCULAR	↗	4
		TOTAL	↗	4,6

Note: 1) The symbol ⌣ formerly denoted flatness.

The symbol ⌒ or — formerly denoted flatness and straightness.

2) Considered "related" features where datums are specified.

3) The symbol || and ◉ formerly denoted parallelism and concentricity, respectively.

4) The symbol ↗ without the qualifier "CIRCULAR" formerly denoted total runout.

5) Where symmetry applies, it is preferred that the position symbol be used.

6) "TOTAL" must be specified under the feature control symbol.

7) Consider the use of position or runout.

Where existing drawings using the above former symbols are continued in use, each former symbol denotes that geometric characteristic which is applicable to the specific type of feature shown.

FIG. 73 GEOMETRIC CHARACTERISTIC SYMBOLS

26

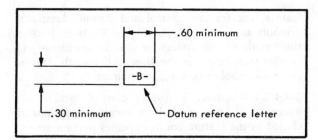

FIG. 74   DATUM IDENTIFYING SYMBOL

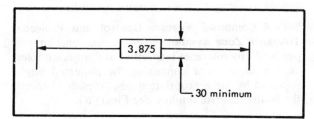

FIG. 75   BASIC DIMENSION SYMBOL

TERM	ABBREVIATION	SYMBOL
Maximum Material Condition	MMC	Ⓜ
Regardless of Feature Size	RFS	Ⓢ
Diameter	DIA	∅
Projected Tolerance Zone	TOL ZONE PROJ	Ⓟ
Reference	REF	(1.250)
Basic	BSC	3.875

FIG. 76   OTHER SYMBOLS

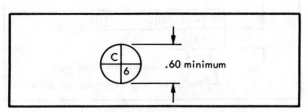

FIG. 77   DATUM TARGET SYMBOL

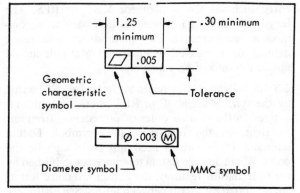

FIG. 78   FEATURE CONTROL SYMBOLS

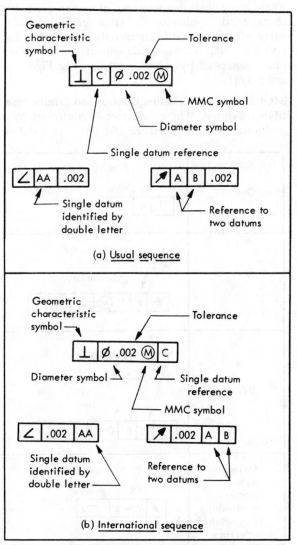

(a) Usual sequence

(b) International sequence

FIG. 79   FEATURE CONTROL SYMBOLS
INCORPORATING DATUM REFERENCES

entry includes the symbol for MMC or RFS. The length of frame is increased as necessary to accommodate requirements. Two methods of referencing datums are illustrated in Figure 79. Methods should not be intermixed on a drawing.

5-3.4.2.1 Each datum reference letter (supplemented by the symbol for MMC or RFS where applicable) is entered in the desired order of precedence, from left to right, in the feature control symbol. Datum reference letter entries need not be in alphabetical order. Where a single datum reference is established by multiple datum features, the datum reference letters are separated by a dash between letters. See Figure 80.

5-3.4.2.2 A composite feature control symbol is used where more than one tolerance of a given geometric characteristic applies to the same feature. A single entry of the geometric characteristic symbol is followed by each tolerance requirement, one above the other, separated by a horizontal line. See Figure 81 and 5-5.6.1.

5-3.4.3 Combined Feature Control and Datum Identifying Symbol. Where a feature is controlled by a positional or form tolerance and serves as a datum

feature, the feature control and datum identifying symbols are combined. See Figure 82. In such cases, the length of the frame for the datum identifying symbol may be either the same as that of the feature control symbol or 0.60 inch minimum.

5-3.4.3.1 Whenever a feature control symbol and datum identifying symbol are combined, datums included in the feature control symbol portion are not considered part of the datum identifying symbol. In the positional tolerance example, Figure 82, a feature is controlled for position (in relation to datums A and B) and identified as datum C. Whenever datum C is referenced elsewhere on the drawing, the reference applies to datum C, not to datums A and B.

5-3.4.4 Combined Feature Control and Projected Tolerance Zone Symbol. Where a positional or perpendicularity tolerance is specified as a projected tolerance zone, a frame containing the projected height followed by the appropriate symbol is placed beneath the feature control symbol. See Figure 83.

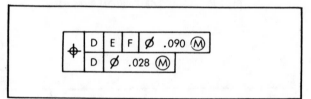

FIG. 81   COMPOSITE FEATURE CONTROL SYMBOL

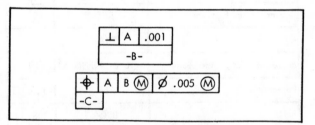

FIG. 82   COMBINED FEATURE CONTROL AND DATUM
IDENTIFYING SYMBOL

One	Primary ⊥ A .005
Two	Primary — Secondary ⊕ B C Ⓜ ⌀ .010 Ⓜ
Three	Primary — Secondary — Tertiary ⊕ F E D ⌀ .014 Ⓜ
Multiple datum features establishing single datum reference	Multiple datum primary ⌀ A – B .002

FIG. 80   ORDER OF PRECEDENCE OF DATUM
REFERENCES

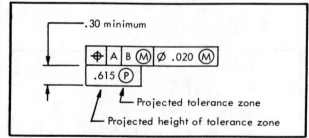

FIG. 83   FEATURE CONTROL SYMBOL WITH A
PROJECTED TOLERANCE ZONE

5-3.5 FEATURE CONTROL SYMBOL PLACEMENT. The symbol is related to the considered feature by one of the following methods (see Figure 84):

(a) Adding the symbol to a note or dimension pertaining to the feature.

(b) Running a leader from the feature to the symbol.

(c) Attaching a side, end, or corner of the symbol frame to an extension line from the feature.

(d) Attaching a side or end of the symbol frame to the dimension line pertaining to the feature.

5-3.6 IDENTIFICATION OF TOLERANCE ZONE. Where the specified tolerance value represents the diameter of a cylindrical zone, the diameter symbol shall be included in the feature control symbol. Where the tolerance zone is other than a diameter, identification is unnecessary, and the specified tolerance value represents the distance between two parallel straight lines or planes, or the distance between two uniform boundaries, as the specific case may be. For roundness, cylindricity, or runout, the tolerance value shown shall represent the full indicator movement. If desired, the term **TOTAL, WIDE ZONE,** or **ON RADIUS** may be used to supplement the tolerance in the appropriate feature control symbol.

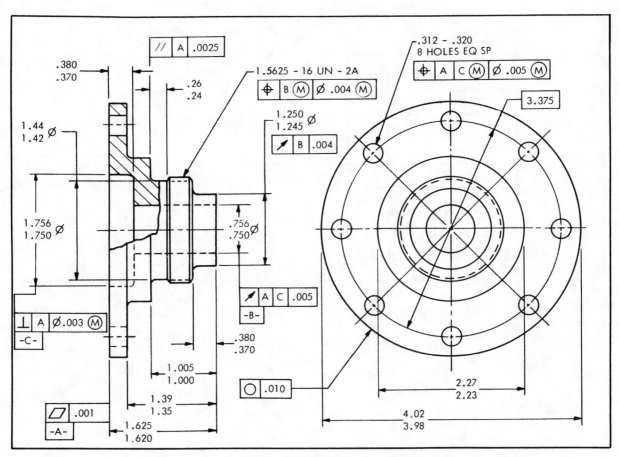

FIG. 84 APPLICATIONS OF GEOMETRIC CHARACTERISTICS SYMBOLS

5-3.7 TABULATED TOLERANCES. Where the tolerance in a feature control symbol is to be tabulated, the letter representing the tolerance is preceded by the abbreviation TOL, as shown in Figure 85.

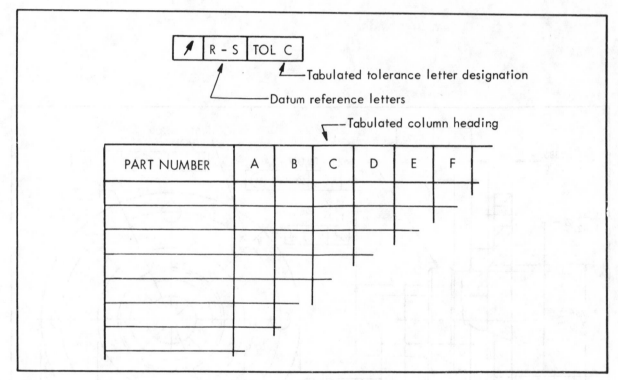

FIG. 85 TABULATED TOLERANCES

5-4 DATUM REFERENCING

5-4.1 GENERAL. This subsection establishes the principle of datum referencing used to relate features of a part to an appropriate datum.

5-4.2 DATUM. A datum indicates one end of a relationship between a toleranced feature and another feature of a part. Features selected as datums are called datum features, while the geometric counterparts with which they are associated are called datum points, lines, planes, cylinders, or axes. See 5-1.6.5.

5-4.3 APPLICATION. As measurements cannot be made from true geometric counterparts, datums are assumed to exist in associated processing equipment. For example, machine tables, surface plates, etc., though not true planes, are of such quality they are used to simulate datum planes for purposes of dimensioning verification.

5-4.3.1 Datum Reference Frame. Surfaces or other features most important to the design of a part are selected as origins for dimensions. Sufficient features are chosen to position the part in relation to a set of three mutually perpendicular planes, jointly called the datum reference frame. All related measurements of the part originate from these planes. See Figure 86. In many instances, a single datum reference frame suffices. Multiple datum reference frames are used where either physical or functional separation of features are such as to require datum reference frames applied at specific locations on the part. To distinguish between datum reference frames, each feature control symbol contains its applicable datum feature references.

5-4.3.2 Datum Features. A feature is selected as a datum based on its geometric relationship to the toleranced feature and the functional requirements of

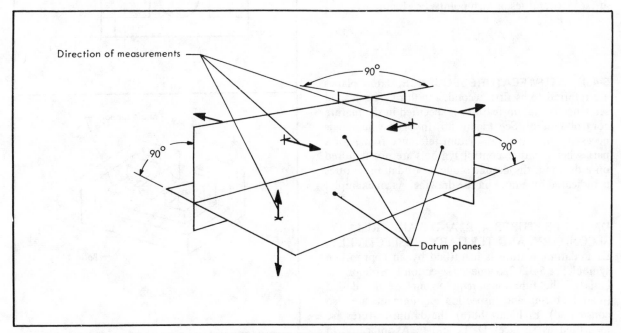

FIG. 86 DATUM REFERENCE FRAME

the design. To ensure proper part interface and assembly, corresponding features of mating parts are used as datums where practicable. Features selected to serve as datums must be distinguishable. In the case of symmetrical parts or parts with identical features, physical identification on the part is recommended. A datum feature must be accessible on the part and of sufficient size to permit subsequent processing operations.

5-4.3.2.1 Selected features of castings, forgings, or weldments may be used temporarily for the establishment of machined surfaces to serve as permanent datum features. Such temporary datum features may or may not be subsequently removed by machining. Permanent datum features should be surfaces or diameters not appreciably changed by machining operations.

5-4.3.2.2 Measurements made from a datum plane do not take into account any variation of the datum surface from the datum plane. Consideration shall be given to the accuracy of datum features relative to design requirements and the degree of control required for related features. If not sufficiently accurate, datum features shall have their variations controlled (as other features are controlled) by specifying form tolerances and surface texture requirements.

5-4.3.2.3 When magnified, flat surfaces of manufactured parts are seen to have geometric irregularities. Contact is made with a datum plane at a number of surface extremities or high points. See Figure 87.

5-4.4 DATUM FEATURE SEQUENCE. Datum planes are referred to as first, second, and third. The desired sequence of datum features is specified in the feature control symbol. See Figure 80. In cases where it is necessary to establish a datum reference frame for a part where feature control symbols are not required on a drawing, the desired sequence of datum features is indicated by a note on the drawing. As an example:

DATUM FEATURES A, B, AND C ARE PRIMARY, SECONDARY, AND TERTIARY, RESPECTIVELY. Each datum feature is identified by an appropriate symbol. See 5-3.3.2. Sequence selection is made on the basis of the functional requirements of the design (manufacturing and inspection requirements are also considered). In Figure 88(a), the datum features are identified as surfaces D, E, and F. As indicated in Figure 88(b), these surfaces are design requirements

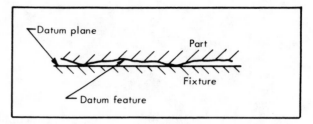

FIG. 87 MAGNIFIED SECTION OF FLAT SURFACE IN CONTACT WITH SIMULATED DATUM PLANE

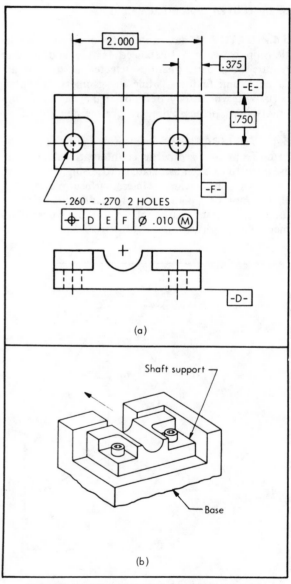

FIG. 88 EXAMPLE OF PART WHERE DATUM FEATURES ARE FLAT SURFACES

necessary for proper assembly and functioning of the part. As surfaces D, E, and F (Figure 88) are the primary, secondary, and tertiary datum features respectively, they appear in that order in the feature control symbol. Depending on the type of tolerance (position or form) and the relationship required, it may be necessary to reference only one or two datum features. The primary datum feature relates the part to the datum reference frame by bringing a minimum of three points into contact with the first datum plane. See Figure 89(a). The part is further related to the frame by bringing at least two points of the secondary datum feature into contact with the second datum plane. See Figure 89(b). The relationship is completed by bringing at least one point of the tertiary datum feature into contact with the third datum plane. See Figure 89(c). As measurements are made from datum planes, positioning of the datum features in relation to these planes ensures a common basis for measurements.

5-4.5 PARTS WITH CYLINDRICAL DATUM FEATURES.
Cylindrical surfaces of parts bear no resemblance to datum planes. However, a cylindrical datum feature is associated with two datum planes represented by centerlines drawn at right angles, intersecting at the axis of the cylindrical datum feature.

5-4.5.1 Three-Plane Relationship. Application of the three plane concept to a cylindrical part is demonstrated in Figure 90(a) and (b).

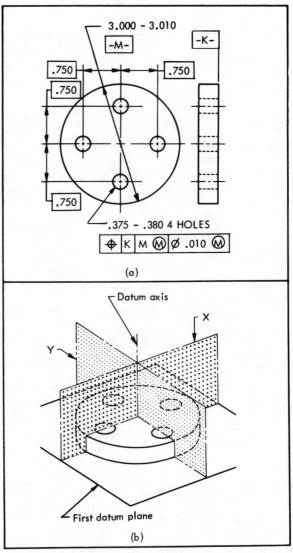

FIG. 90 EXAMPLE OF PART WITH CYLINDRICAL DATUM FEATURE

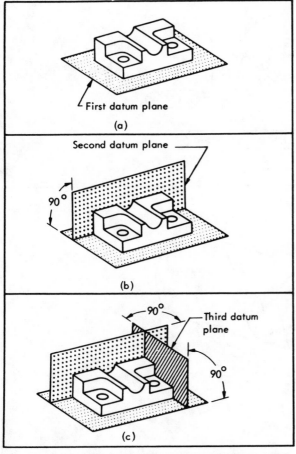

FIG. 89 SEQUENCE OF DATUM FEATURES RELATES PART TO DATUM REFERENCE FRAME

NOTE: In Figure 90(a), datum feature K is a sealing surface to which the bolt clearance holes are normal. For design purposes it has been selected as the primary datum feature. The holes are related to cylindrical datum feature M and are dimensioned from centerlines through the center of M. As the part has depth, the centerlines represent planes through the part center. See Figure 90(b). These planes represent the second and third datum planes (X and Y) respectively. The clearance holes are located relative to these three datum planes.

In cases where a diameter is designated as a datum feature, the line formed by the intersection of two planes is the datum axis.

NOTE: In Figure 90, the sequence of planes X and Y is immaterial as rotation of the pattern about the datum axis has no effect on the function of the part. The feature control symbol references two datum features: (1) K, the primary datum feature—a flat surface associated with the first datum plane; and, (2) M, the secondary datum feature (cylindrical) associated with the second and third datum planes (or datum axis).

5-4.5.2 Angular Orientation. To establish angular orientation of two planes about the datum axis, a third datum feature is referenced in the feature control symbol. See Figure 91.

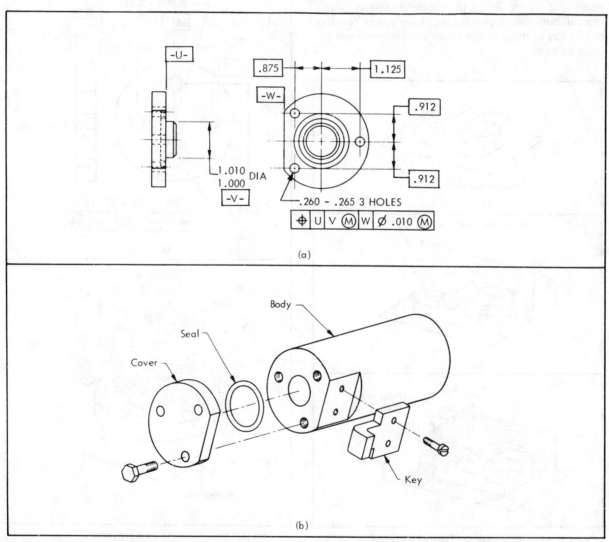

FIG. 91   EXAMPLE OF PART WHERE ANGULAR ORIENTATION IS IMPORTANT

NOTE: In Figure 91, surface U is in contact with its mating surface and the clearance holes are normal to it; therefore, it is the primary datum feature. The secondary datum feature is diameter V of the cover, which provides axial alignment with the body. As V is a diameter, it is associated with two datum planes. The flats of both cover and body align to provide for mounting of the key. Accordingly, the angular orientation of the datum planes through V is established in relation to an auxiliary plane associated with datum feature W and perpendicular to the first datum plane.

5-4.6 DATUM FEATURES SUBJECT TO SIZE VARIATIONS.
Datum features such as diameters and widths differ from flat surfaces in two ways: (1) Dimensions originate from centerlines or center planes rather than from the surface of the feature. Example: Diameter A of the part in Figure 92(a) is a datum feature, and measurements indicated by the dimensions originate from datum planes which intersect at a datum axis. (2) Diameters and widths are subject to variations in size as well as surface irregularities.

NOTE: As cylindrical datum features vary in size, the application of RFS or MMC is considered in the definition, such as diameter A in Figure 92(a). As a flat surface is associated with its geometric counterpart (a true plane), so is a diameter associated with its geometric counterpart, a true cylinder. The datum axis (the intersection of two datum planes) is the axis of the true cylinder associated with a cylindrical datum feature.

5-4.6.1 Primary Datum Feature, RFS.
The sequence of datum references in a feature control symbol has a direct bearing on the function of the part. In Figure 92 (b), diameter A is the primary datum feature, RFS is applied, and surface B is the secondary datum feature. A is specified first in the feature control symbol. The datum axis is the axis of the smallest true cylinder which contacts the high points of surface A.

NOTE: The smallest true cylinder is simulated by processing equipment with centering devices (chuck, mandrel, etc.).

The first and second datum planes intersect at the datum axis and at least one point of surface B contacts the third datum plane to complete the three-plane relationship.

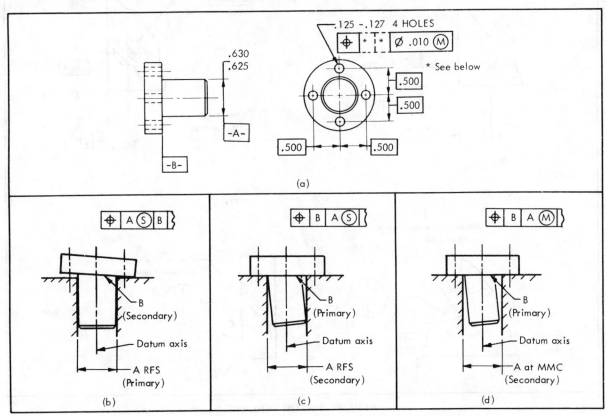

FIG. 92 EFFECT OF RFS OR MMC APPLIED TO DATUM FEATURES

5-4.6.2 Secondary Datum Feature, RFS. In Figure 92(c), B is the primary datum feature, A is the secondary, and RFS is applied. The datum axis is the axis of the smallest true cylinder perpendicular to the first datum plane (the plane which contacts at least three high points of surface B) that will contact points of surface A. The second and third datum planes intersect at the datum axis to complete the three-plane relationship.

5-4.6.3 Secondary Datum Feature at MMC. In Figure 92(d), B is the primary datum feature, A is the secondary, and MMC is applied to datum feature A. The datum axis is the axis of a true cylinder of MMC size perpendicular to the first datum plane (the plane which contacts at least three high points of surface B).

NOTE: The true cylinder is simulated in the processing equipment by a diameter of fixed (MMC) size.

The second and third datum planes intersect at the datum axis to complete the three-plane relationship.

5-4.7 DATUM TARGETS. Datum targets define points, lines or areas on a part used in establishing datum planes. To apply the three-plane concept to certain parts, specific points, lines or areas on features are designated to establish datum planes. Examples of such features are non-planar or uneven surfaces produced by casting, forging or molding; surfaces adjacent to welding; thin section surfaces subject to

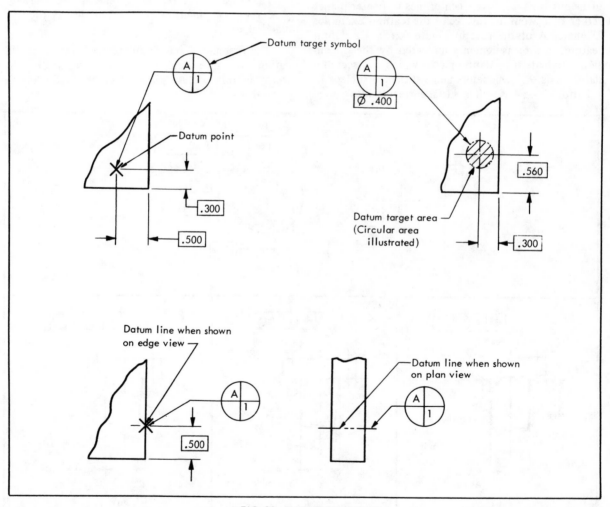

FIG. 93   DATUM TARGETS

bowing, warping or other inherent or induced distortion. These surfaces may contain variations in form to such a degree that use of the entire surface as a datum feature may not satisfy the engineering intent.

NOTE: A nominally flat cast surface placed in contact with a surface plate simulating a datum plane may actually rock, creating uncertainty of measurements. Non-planar or uneven surfaces may also cause differences in measurement between manufacturing and verification fixtures if the location features are not duplicated in position, shape and size.

5-4.7.1 Datum Target Dimensioning. Datum targets are dimensioned on an engineering drawing using basic or toleranced dimensions. Dimensions locating datum targets in a set should be dimensionally related to each other or have a common origin.

5-4.7.2 Datum Target Points, Lines or Areas. The method of designating datum points, lines or areas is shown in Figure 93. A datum area may be any defined shape and identified by diagonal slash lines within a phantom outline. The size of the target controls the area of flat contact necessary to assure establishment of the datum. See 5-3.3.8 for the datum target symbol.

5-4.7.3 Datum Target Application. Adequate views to depict the datum targets must be provided. A representation of datum targets on a part is shown in Figure 94. Datum targets may vary from the conventional three point, two point, one point target orientation to assure a fixed position of the part. Datum targets, while intended for use on parts having irregular surfaces and geometric contours (such as castings, forgings, complex sheet metal shapes and airfoils), may be applied to any part where design requirements dictate use of specific datum points, lines or areas on surfaces rather than use of entire surfaces as datums.

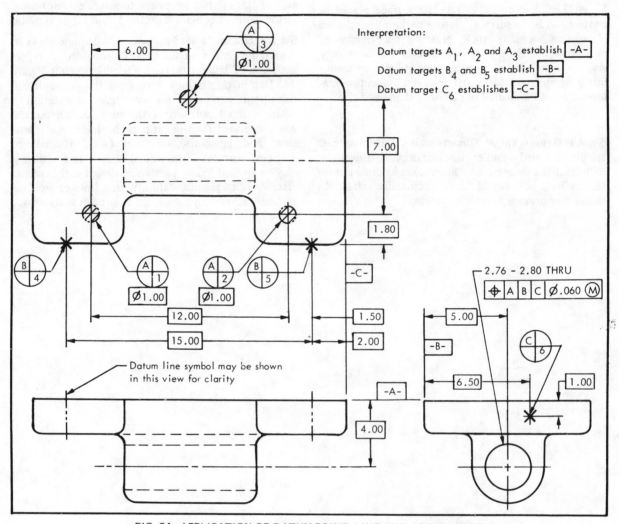

Interpretation:
Datum targets A_1, A_2 and A_3 establish -A-
Datum targets B_4 and B_5 establish -B-
Datum target C_6 establishes -C-

FIG. 94   APPLICATION OF DATUM POINT, LINE AND AREA TARGETS

5-5 TOLERANCES OF LOCATION

5-5.1 GENERAL. This subsection establishes the principles of tolerances of location. Included are position, concentricity, and symmetry used to control the following relationships:

(a) Center distance between such features as holes, slots, bosses, and tabs.

(b) Location of features (such as above), as a group, from datum features such as plane and cylindrical surfaces.

(c) Coaxiality between features.

(d) Features with center distances equally disposed about a datum axis or plane.

5-5.2 POSITIONAL TOLERANCING. A positional tolerance defines a zone within which the axis or center plane of a feature is permitted to vary from true (theoretically exact) position. Basic dimensions establish the true position from specified datum features and between interrelated features. Applicable positional tolerances are placed in feature control symbol blocks.

5-5.3 ADVANTAGES OF POSITIONAL TOLERANCING. See Appendix B.

5-5.4 SPECIFYING POSITIONAL TOLERANCES. Basic dimensions are used to establish the true position of a feature. The permissible variation is expressed as a positional tolerance.

5-5.4.1 True Position. The term "true position" denotes the theoretically exact position of a feature.

5-5.4.1.1 Positional tolerancing is identified by a characteristic symbol directed to a feature. See Figure 95.

5-5.4.2 Method. The following paragraphs describe methods used in expressing positional tolerances.

5-5.4.2.1 The location of each feature (hole, slot, stud, etc.) is given by untoleranced basic dimensions. Many drawings are based on a schedule of general tolerances usually provided near the drawing title block. See American National Standard for Drawing Sheet Size and Format, Y14.1-1957. On such drawings, dimensions locating true position must be excluded from the general tolerance in one of the following ways:

(a) Applying the basic dimension symbol to each of the basic dimensions. See Figure 95.

(b) Applying the word BASIC (or BSC) to each of the locating dimensions. See Figure 95.

(c) Specifying on the drawing the general note: UNTOLERANCED DIM LOCATING TP ARE BASIC. See Figure 96.

5-5.4.2.2 A feature control symbol is added to the note used to specify the size and number of features. See Figures 95 thru 98. These figures show common types of feature pattern dimensioning.

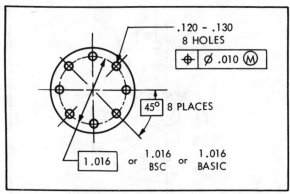

FIG. 95 IDENTIFYING BASIC DIMENSIONS

5-5.4.2.3 Identify datum features on drawings in those instances where failure to so identify could result in the wrong selection of intended features. For example, in Figure 99, if datum references had been omitted, it would not be clear whether the inside diameter (Surface B) or the outside diameter was the intended datum for the true position dimensions. For datum references showing order of precedence, see 5-3.4.2.1. Methods for indicating datums are as follows (listed in order of preference):

(a) Identifying intended features with a datum symbol and including the applicable datum references in the feature control symbol.

(b) Implication, see Figure 111.

5-5.4.3 Application to Base Line and Chain Dimensioning. True position dimensioning can be applied as base line dimensioning or as chain dimensioning. See Figures 100 and 101. Since basic (true position) dimensions are used, and assuming the same positional tolerance, the resultant tolerance between any two holes will be the same for base line dimensioning as for chain dimensioning. This also applies to angular dimensions, whether base line or chain type.

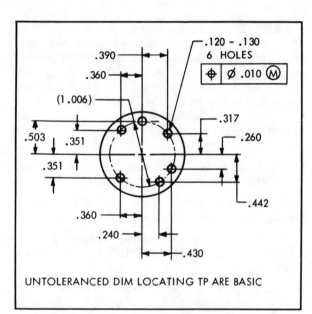

FIG. 96   IDENTIFYING BASIC DIMENSIONS

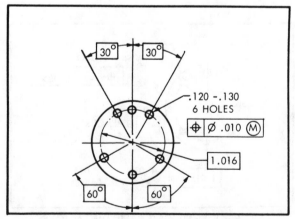

FIG. 97   POSITIONAL TOLERANCING

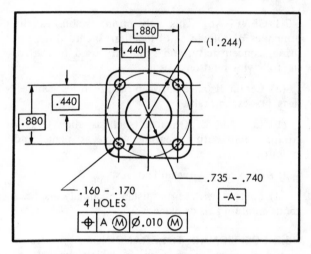

FIG. 98   POSITIONAL TOLERANCING

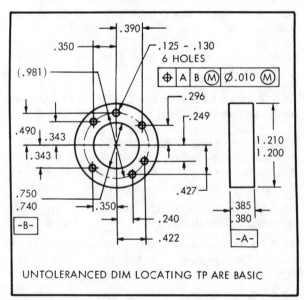

FIG. 99   POSITIONAL TOLERANCING WITH DATUM
REFERENCES

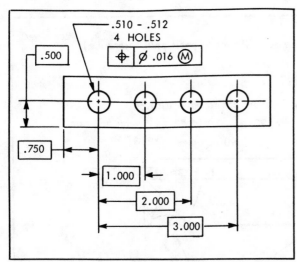

FIG. 100 TRUE POSITIONS DIMENSIONED USING BASE
LINE DIMENSIONING

5-5.4.4 Positional Tolerancing—MMC. Positional tolerancing for features of size is applied either on an MMC basis or on an RFS basis. See 5-5.5. Where the MMC basis is intended see 5-2.12.

5-5.4.5 Positional Tolerancing—RFS. RFS is applied where functional requirements do not allow the application of MMC. In instances where it is necessary to apply RFS, the datum reference, as applicable, and the specified tolerance is followed by the RFS symbol. See Figure 105. See Rules #2 and #2a in 5-2.12.

5-5.5 FUNDAMENTAL EXPLANATION OF POSITIONAL TOLERANCING. The following explanation is basic to positional tolerancing. Other interpretations—those that apply to special cases—will be found at suitable points further on in this subsection.

5-5.5.1 The Meaning of MMC as Related to Positional Tolerancing. MMC by itself means a feature of a finished product contains the maximum amount of material permitted by the toleranced size dimensions for that feature. Thus, for holes, slots, and other internal features, maximum material is the condition where these features are at their minimum allowable sizes. For shafts, as well as for bosses, lugs, tabs, and other external features, maximum material is the condition where these are at their maximum allowable sizes.

5-5.5.1.1 The positional tolerance and maximum material condition of mating features are considered in relation to each other.

5-5.5.1.2 Where a positional tolerance applies at MMC, the explanation is as follows:

(a) *In Terms of the Cylindrical Surface of a Hole.* While maintaining the specified diameter limits of the hole, no element of the hole surface shall be inside a theoretical cylinder, located at true position, having a diameter equal to that of Cylinder A of Figure 102.

(b) *In Terms of the Axis of the Hole.* Where a hole is at MMC (minimum diameter), its axis must fall within a cylindrical tolerance zone whose axis is located at true position, having a diameter equal to the positional tolerance. See Figure 103(a) and (b). This tolerance zone also defines the limits of variation in the attitude of the axis of the hole in relation to the datum surface. See Figure 103(c). It is only when the feature is at MMC that the specified positional tolerance applies. Where the actual size of the feature is at

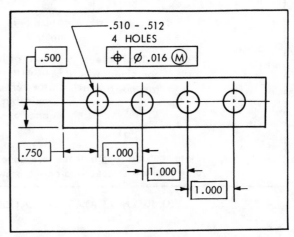

FIG. 101 TRUE POSITIONS DIMENSIONED USING
CHAIN DIMENSIONING

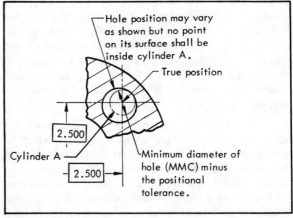

FIG. 102 TOLERANCE ZONE FOR SURFACE OF HOLE
AT MMC

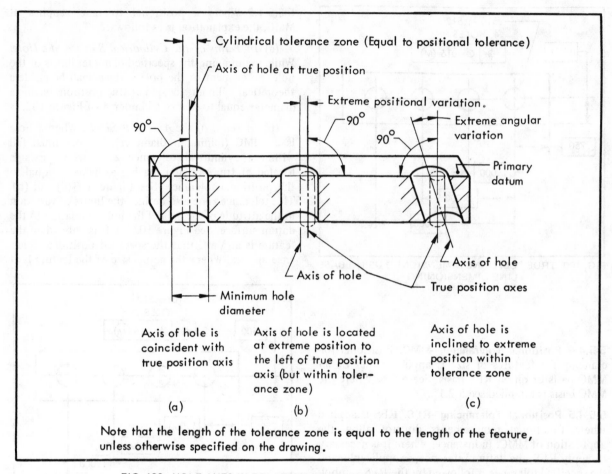

FIG. 103 HOLE AXES IN RELATION TO POSITIONAL TOLERANCE ZONES

other than MMC, additional positional tolerance results. See Figure 104. This increase of positional tolerance is equal to the difference between the specified maximum material limit of size (MMC) and the actual finished size of the feature. The specified positional tolerance for a feature may be exceeded where the size is not at MMC and still satisfy function and interchangeability requirements.

5-5.5.1.3 In many instances a group of features (such as a group of mounting holes in a flange) must be positioned relative to an integral datum at MMC. See Figure 98. Where the datum is at MMC, the axis of the datum feature determines the location of the pattern of features as a group. Where the datum departs from MMC, the pattern of features (as a group) is permitted to be displaced about this axis. This displacement is equal to one-half the difference between the MMC size

of the datum feature and its actual size, applied in any direction. It permits a shift of the pattern of features (as a group) relative to the datum axis without affecting the positional tolerance of the features relative to each other.

5-5.5.2 Zero Positional Tolerance at MMC. The concept of positional tolerance at MMC as described in the preceding paragraphs can be extended in certain applications where it is desired to provide a greater tolerance within functional limits than would otherwise be allowed. This is accomplished by adjusting the minimum limit of hole size to the absolute minimum required for insertion of an applicable fastener located precisely at true position and specifying a zero positional tolerance at MMC. See B6, Appendix B. In this case, the positional tolerance

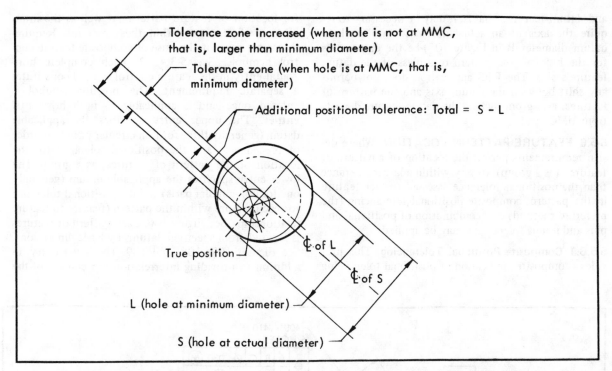

FIG. 104   INCREASE IN POSITIONAL TOLERANCE WHERE HOLE IS NOT AT MMC

allowed is totally dependent on the actual size of the considered feature, as explained in 5-2.12.3. Zero positional tolerancing at MMC should be used with caution since the adjusted size tolerance is larger and requires a judicious selection (by the drawing user) of an actual clearance hole size, which determines the allowable positional tolerance for its location.

5-5.5.3 The Explanation of RFS as Related to Positional Tolerancing. RFS, where applied to the positional tolerance of circular features, requires the axis of each feature be located within the specified positional tolerance regardless of the size of the feature.

5-5.5.3.1 In Figure 105, the six holes may vary in size from 0.9994 to 1.0000. The six individual hole diameters might measure 0.9995, 0.9996, 0.9997, 0.9998, 0.9999 and 1.0000. In order to minimize spacing errors each hole must be located within the specified positional tolerance regardless of the size of that hole. A hole at LMC (1.0000 diameter) is as accurately located as a hole at MMC (0.9994 diameter). This positional control is more restrictive than the MMC concept more commonly applied to positional tolerancing.

5-5.5.3.2 The functional requirements of some designs may require RFS be applied to both the hole pattern

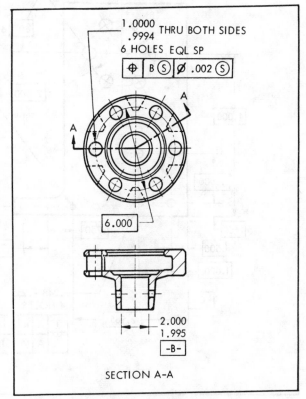

FIG. 105   RFS APPLIED TO A FEATURE AND DATUM

and datum feature. That is, it may be necessary to require the axis of an actual datum feature (such as datum diameter B in Figure 105) be the datum axis for the holes in the pattern regardless of the datum feature's size. The RFS application does not permit any shift between the datum axis and the pattern of features, as a group, where the datum feature departs from MMC.

5-5.6 FEATURE PATTERN LOCATION. Where design requirements permit the location of a pattern of features (as a group) to vary within a larger tolerance than the positional tolerance assessed to each feature in the pattern, composite positional tolerancing (the preferred method) or a combination of positional and plus and minus tolerancing may be applied.

5-5.6.1 Composite Positional Tolerancing. This provides a composite application of positional tolerancing

for location of feature patterns as well as the interrelation of features within these patterns. Requirements are annotated by use of composite feature control symbols. See 5-3.4.2.2. Each complete horizontal entry in the feature control symbol constitutes a separate requirement. The position symbol is entered once and is applicable to both horizontal entries. The upper entry specifies the applicable datum (generally three for non-circular parts), in order of precedence, and the positional tolerance for the location of the pattern of features, as a group. The lower entry specifies the applicable datum (generally one for non-circular parts) and the positional tolerance for each feature within the pattern (feature-to-feature relationship). See Figure 106. Each pattern of features is located from specified datums by basic dimensions. See Figures 107, 108, and 109. The lower entry, in addition to providing interrelationship control of the

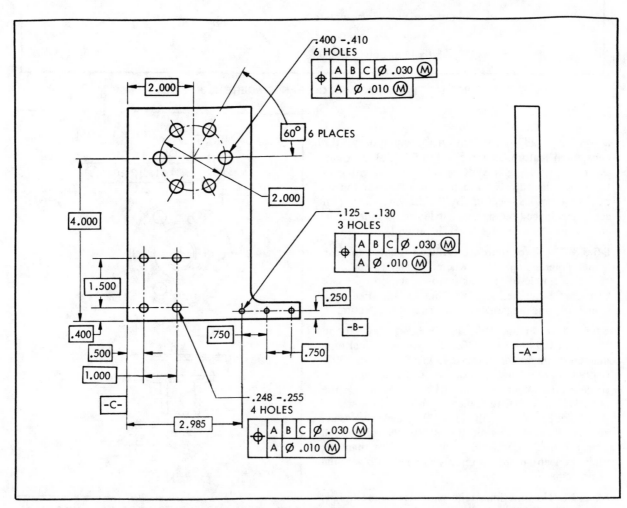

FIG. 106   HOLE PATTERNS LOCATED BY COMPOSITE POSITIONAL TOLERANCING, SPECIFIED DATUMS

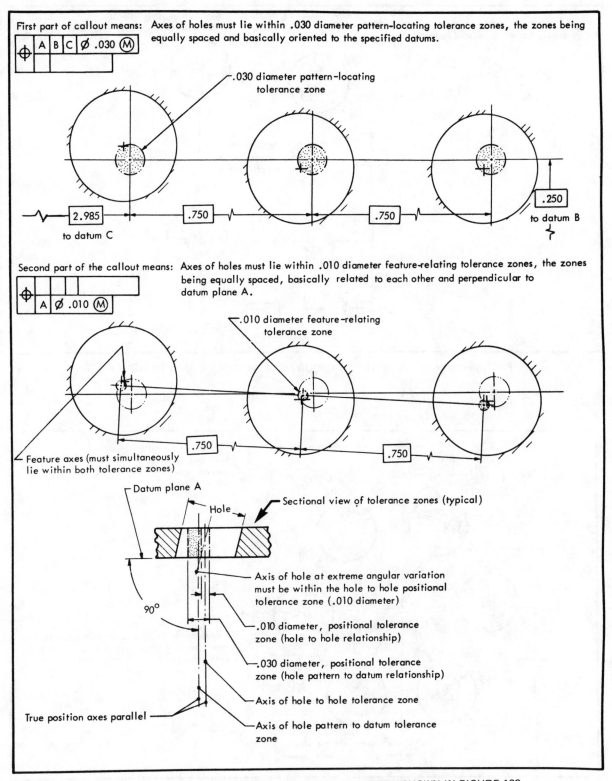

First part of callout means: Axes of holes must lie within .030 diameter pattern-locating tolerance zones, the zones being equally spaced and basically oriented to the specified datums.

Second part of the callout means: Axes of holes must lie within .010 diameter feature-relating tolerance zones, the zones being equally spaced, basically related to each other and perpendicular to datum plane A.

FIG. 107   TOLERANCE ZONES FOR THREE-HOLE PATTERN SHOWN IN FIGURE 106

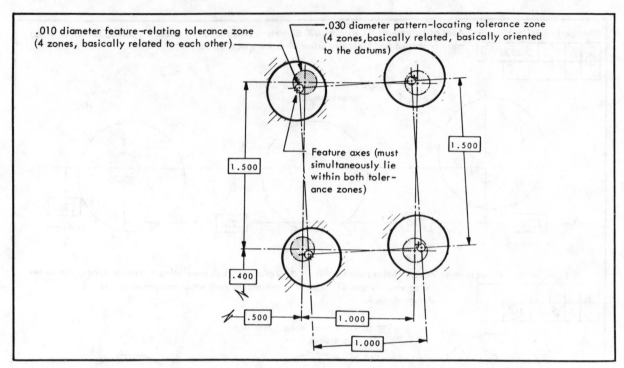

FIG. 108   TOLERANCE ZONES FOR FOUR-HOLE PATTERN SHOWN IN FIGURE 106

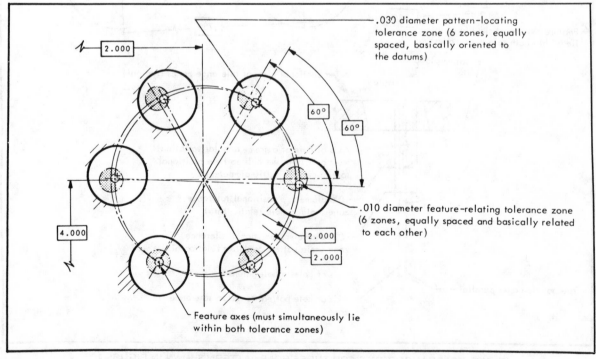

FIG. 109   TOLERANCE ZONES FOR SIX-HOLE PATTERN SHOWN IN FIGURE 106

features in each pattern, controls the extent of angular variation (perpendicularity in the case of Figure 106) of each feature's axis in relation to the plane established by datum **A**. As can be seen from the sectional view of the tolerance zones in Figure 107, the axes of both the large and small zones are parallel. The axes of the holes may vary obliquely only within the confines of the respective smaller positional tolerance zones. The axes of the holes must lie within the larger tolerance zones and also within the smaller tolerance zones. In certain instances, a portion of the smaller zones may fall beyond the peripheries of the larger tolerance zones. However, this portion of the smaller tolerance zone is not usable because the axis of the feature must not violate the larger tolerance zone. See Figure 107.

NOTE: The zones in these figures are shown as they exist at MMC of the features depicted in Figure 106. The large zones would increase in size by the amount the features depart from MMC, as would the smaller zones; the two zones are not cumulative. Composite positional tolerancing may be applied to patterns of features on circular parts. See Figure 110.

5-5.6.2 Combination of Positional and Plus and Minus Tolerancing. Where this combination of tolerancing methods is used, the plus and minus dimensions locating the pattern do not state a datum relationship. Their origins are implied by the surfaces from which the dimensions originate (implied datums). The plus and minus tolerances assigned to the coordinate dimensions provide zones within which the center (true position) of each positional tolerance zone must lie. The axis of each feature can lie outside its respective coordinate tolerance zone by an amount equal to one-half the specified positional tolerance when the feature is at MMC. Typical examples of tolerance zones are as follows:

(a) Figure 111 shows an example of three different hole patterns where a combination of tolerancing methods is used. The resultant tolerance zones for each pattern are shown in Figures 112, 113 and 114. In any one of the three patterns, the center (true position) of each positional tolerance zone may deviate from the mean position in relation to the implied datum surfaces by an amount equal to 0.030/2 or 0.015. Within the respective pattern, the axis of each hole may deviate an additional amount equal to 0.010/2 or 0.005. If the positional tolerance is applied on an MMC basis (as it is in Figure 111), any hole not at its MMC is permitted an increase of positional tolerance. See Figure 104. Therefore, in addition to a shift of each pattern (as a group) within their respective plus and minus tolerance zones, the axis of each hole may individually shift outside these zones by an amount equal to one-half the positional tolerance

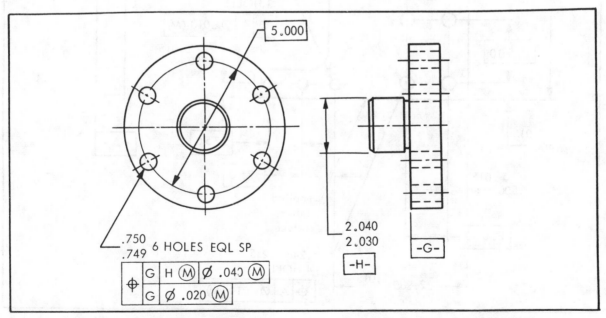

FIG. 110 COMPOSITE POSITIONAL TOLERANCING OF A CIRCULAR PATTERN OF FEATURES

allowed. Figure 112 shows a typical sectional view of the tolerance zones. As can be seen from this view, the axes of the coordinate and positional tolerance zones are parallel. However, the axis of the hole may vary obliquely as long as it lies within the positional tolerance zone.

(b) Figure 115 shows an example of a hole pattern where the basic plane of the pattern is located from an implied datum face by a dimension with a plus and minus tolerance. The same explanation given in (a) also applies to Figure 115. At MMC, the hole pattern (as a group) is permitted to deviate from mean

position in relation to the implied datum by an amount equal to ± 0.030, and each hole may deviate an additional 0.010 at MMC. The total deviation between any two holes would be 0.010 × 2 or 0.020.

5-5.6.3 Multiple Patterns of Features Located by Basic Dimensions Relative to Common Datums. Multiple patterns of features located by basic dimensions from common datum features not subject to size tolerances are considered a single, composite pattern. For instance, in the two pattern example in Figure 116, interrelationship between the patterns is maintained

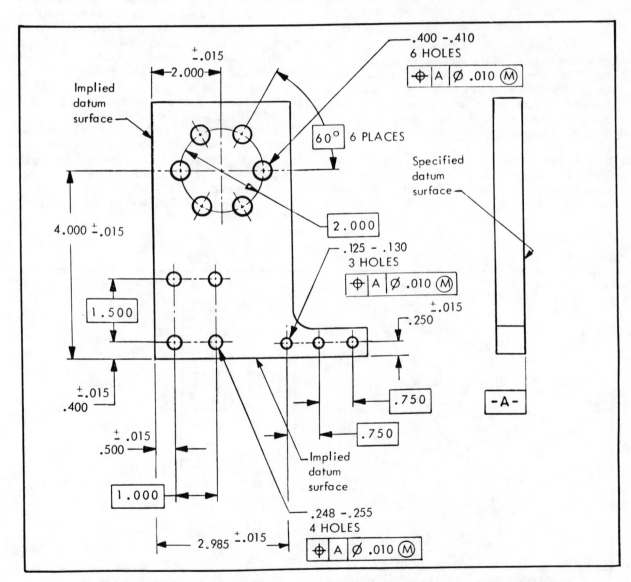

FIG. 111   HOLE PATTERNS LOCATED BY PLUS AND MINUS TOLERANCING, IMPLIED DATUMS

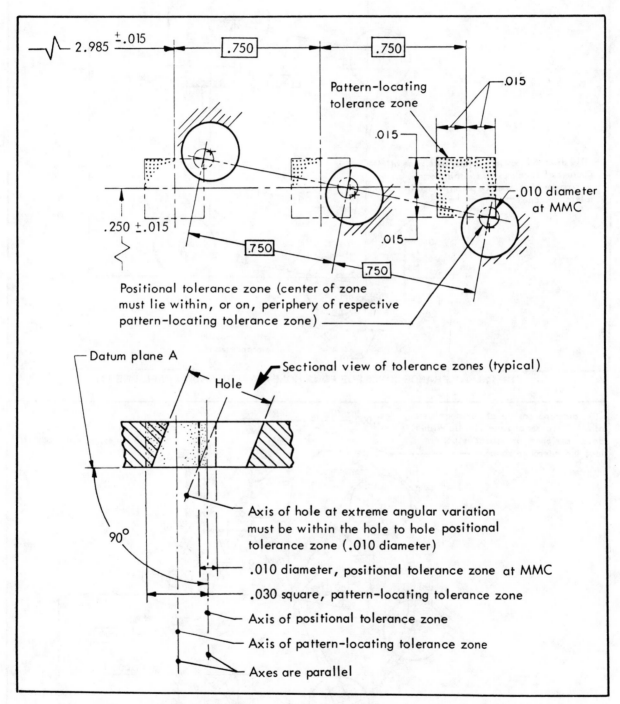

FIG. 112   TOLERANCE ZONES FOR THREE-HOLE PATTERN SHOWN IN FIGURE 111

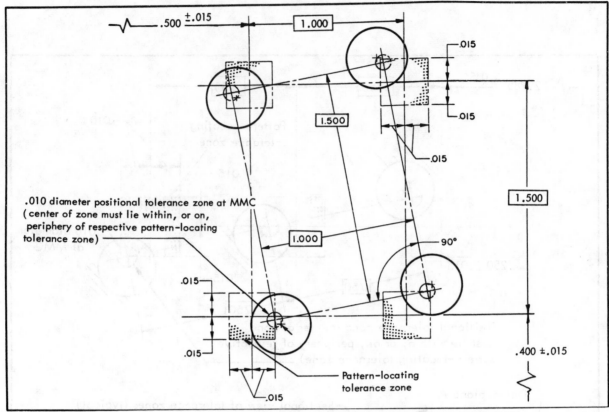

FIG. 113 TOLERANCE ZONES FOR FOUR-HOLE PATTERN SHOWN IN FIGURE 111

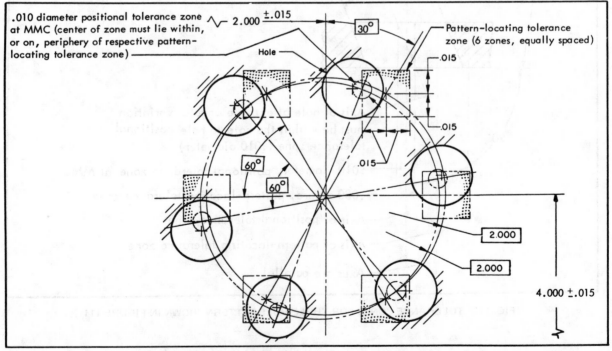

FIG. 114 TOLERANCE ZONE FOR SIX-HOLE PATTERN SHOWN IN FIGURE 111

within their respective positional tolerance zones as shown in Figure 117. Multiple patterns of features located by basic dimensions from common datum features that are subject to size tolerances are also considered a single composite pattern if their respective feature control symbols contain the same datums in the same order of precedence and at the same size

consideration. If such interrelationship is not required between one pattern and any other pattern (or patterns), a notation such as SEP REQT is placed beneath each applicable feature control symbol. See Figure 118. This allows each feature pattern (as a group) to shift independently of each other relative to the datum axis and denotes an independent relationship between the patterns.

5-5.7 PROJECTED TOLERANCE ZONE CONCEPT.
Application of this concept is recommended where the variation in perpendicularity of threaded holes (as in Figure 119) or press fit holes could cause studs, bolts, or pins to interfere with mating parts. Figure 120 illustrates how the projected tolerance zone concept realistically treats the condition shown in Figure 119. Note that it is the variation in perpendicularity of the portion of the bolt passing through the mating part that is significant. The location and perpendicularity of the threaded hole is only of importance insofar as it affects the extended portion of the engaging bolt.

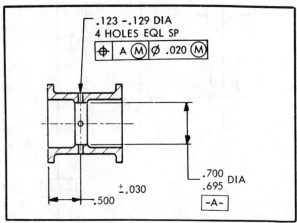

FIG. 115 RADIAL HOLE PATTERN LOCATED BY PLUS
AND MINUS TOLERANCE

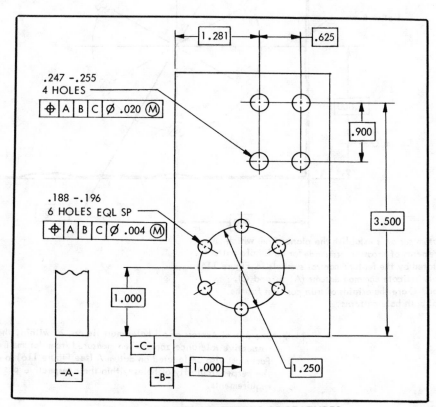

FIG. 116 MULTIPLE PATTERNS OF FEATURES

51

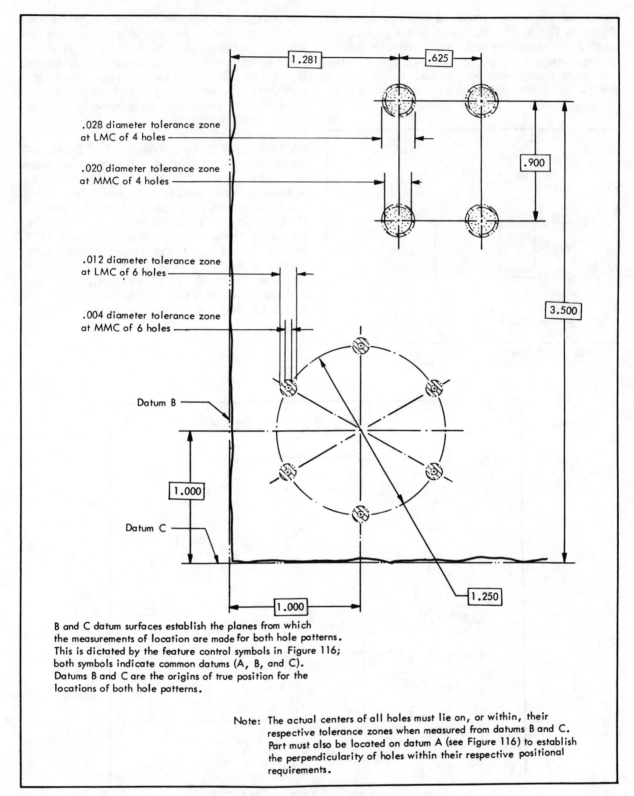

.028 diameter tolerance zone
at LMC of 4 holes

.020 diameter tolerance zone
at MMC of 4 holes

.012 diameter tolerance zone
at LMC of 6 holes

.004 diameter tolerance zone
at MMC of 6 holes

Datum B

Datum C

B and C datum surfaces establish the planes from which
the measurements of location are made for both hole patterns.
This is dictated by the feature control symbols in Figure 116;
both symbols indicate common datums (A, B, and C).
Datums B and C are the origins of true position for the
locations of both hole patterns.

Note: The actual centers of all holes must lie on, or within, their
respective tolerance zones when measured from datums B and C.
Part must also be located on datum A (see Figure 116) to establish
the perpendicularity of holes within their respective positional
requirements.

FIG. 117   TOLERANCE ZONES FOR PATTERNS SHOWN IN FIGURE 116

Figure 121 illustrates an application of positional tolerances for threaded holes with a projected tolerance zone specified. Specified values of projected tolerance zone heights are minimums, and the specified value

represents the maximum permissible mating part thickness.

NOTE: For through holes, the desired surface

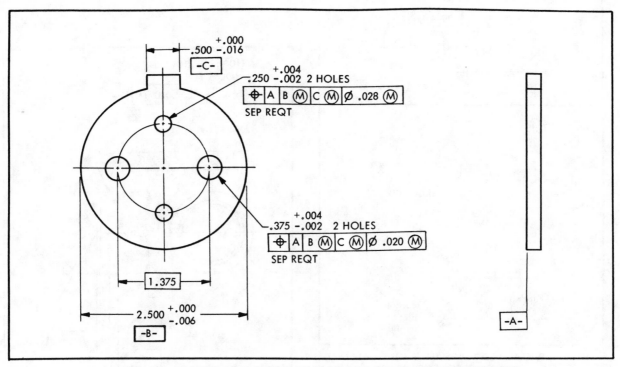

FIG. 118 MULTIPLE PATTERNS OF FEATURES, SEPARATE REQUIREMENTS

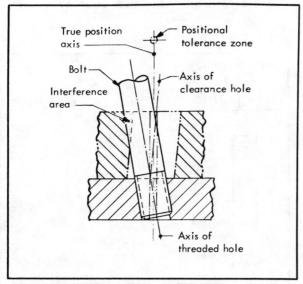

FIG. 119 INTERFERENCE DIAGRAM, FASTENER
AND HOLE

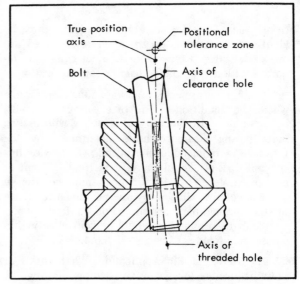

FIG. 120 BASIS FOR PROJECTED TOLERANCE ZONE

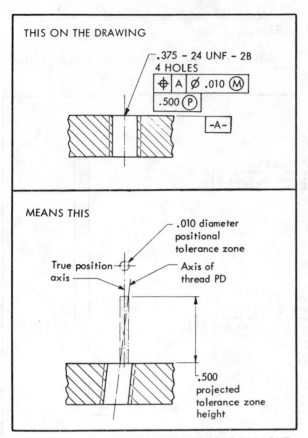

FIG. 121   PROJECTED TOLERANCE ZONE CALLOUT
AND INTERPRETATION

after installation, and the specification of a projected tolerance zone is unnecessary. A projected tolerance zone is applicable where threaded or plain holes for studs or pins are located on a detail part drawing. In

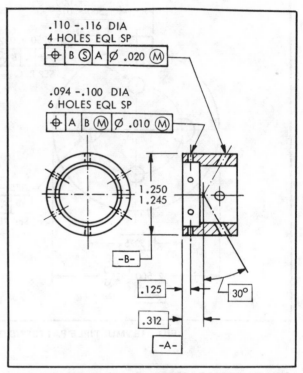

FIG. 122   NONPARALLEL HOLES LOCATED AT AXIS
OF OUTSIDE DIAMETER

from which the tolerance zone is projected must be identified. Failure to do so could result in an improper selection. Complex or unusual cases may require addition of a note or diagram for clarification.

Specifying a projected tolerance zone is applicable where the condition is one of ensuring that studs, bolts, or press fit pins do not interfere with mating parts having clearance holes determined by the formulas recommended in Appendix C. Use of this concept does not require any further enlargement of clearance holes to provide for the extreme variation in perpendicularity allowed by the positional tolerance of the mating part features. For controlling perpendicularity closer than would be controlled within a positional tolerance zone, see 5-6.6.4.

5-5.7.1 Stud and Pin Application. Where studs or press fit pins are located on an assembly drawing, the specified positional tolerance applies only to the height of the projecting portion of the stud or pin

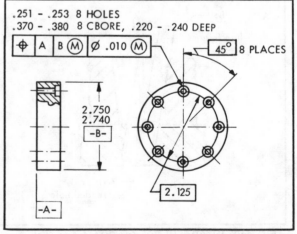

FIG. 123   SAME POSITIONAL TOLERANCE FOR HOLES
AND COUNTERBORES, SAME DATUM REFERENCES

these cases, the specified projected height should equal the maximum permissible height of the stud or pin after installation, not the mating part thickness.

5-5.8 NONPARALLEL HOLES. Positional tolerancing lends itself to patterns of holes where axes are not parallel to each other. See Figure 122.

5-5.9 COUNTERBORED HOLES. Where positional tolerances are used to locate coaxial features, such as counterbored holes, the following practices apply, as applicable:

(a) Where the same positional tolerance is used to locate both holes and counterbores, a single feature control symbol is placed under the notes specifying hole and counterbore requirements. See Figure 123. The positional tolerance applies simultaneously for holes and counterbores.

(b) Where different positional tolerances are used to locate holes and counterbores (relative to common datum features), two feature control symbols are used. One feature control symbol is placed under the note specifying hole requirements and the other under the note specifying counterbore requirements. See Figure 124. The positional tolerances apply simultaneously for holes and counterbores. See 5-5.6.3.

(c) Where positional tolerances are used to locate holes and also control individual counterbore-to-hole relationships (relative to different datum features), two feature control symbols are used, as in (b) above. In addition, a note is placed under the datum identifying symbol for the hole and the feature control symbol for the counterbore, indicating the number of places each applies on an individual basis. See Figure 125.

5-5.10 CLOSER CONTROL AT ONE END OF A FEATURE. Where design permits, different positional tolerances may be specified for the extremities of long holes; this establishes a conical rather than a cylindrical tolerance zone. See Figures 126 and 127.

5-5.11 NONCIRCULAR FEATURES. The basic principles of true position dimensioning for circular features, such as holes and bosses, apply also to noncircular features, such as open end slots, tabs and elongated holes. For such features, the positional tolerance is applied only to surfaces related to the center

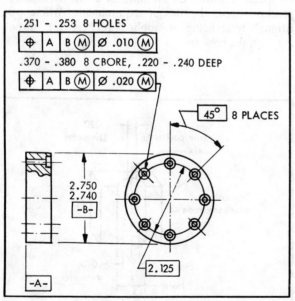

FIG. 124 DIFFERENT POSITIONAL TOLERANCES FOR HOLES AND COUNTERBORES, SAME DATUM REFERENCES

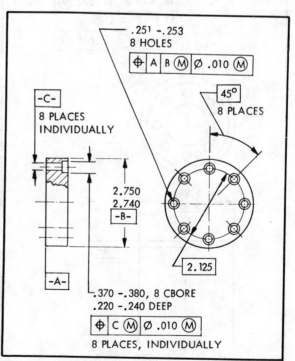

FIG. 125 POSITIONAL TOLERANCES FOR HOLES AND COUNTERBORES, DIFFERENT DATUM REFERENCES

plane of the feature. The tolerance value represents a distance between two parallel planes. The diameter symbol is omitted from the feature control symbol. See Figures 128 and 129.

5-5.11.1 Noncircular Features at MMC. Where a positional tolerance of a noncircular feature applies at MMC, the following apply:

(a) *In Terms of the Surfaces of a Slot.* While maintaining the specified width limits of the slot, no element of the side surfaces of the slot shall be inside a theoretical boundary defined by two parallel planes equally disposed about true position and separated by a distance equal to that shown for Zone W. See Figure 130.

(b) *In Terms of the Center Plane of a Slot.* Where a slot is at MMC, its center plane must be within a tolerance zone defined by two parallel planes equally disposed about true position, having a width equal to the

positional tolerance. See Figure 131. This tolerance zone also defines the limits within which variations in attitude of the center plane of the slot must be confined.

(c) *In Terms of the Boundary for an Elongated Hole.* While maintaining the specified size limits of the elongated hole, no element of its surface shall be inside a theoretical boundary of identical shape located at true position. The size of the boundary is equal to the MMC size of the elongated hole minus its positional tolerance. See Figure 132. In this example, a greater positional tolerance is allowed for its length than its width. Where the same positional tolerance can be allowed for both, only one feature control symbol is necessary, directed to the feature by a leader and separated from the size dimension.

5-5.12 COAXIALITY CONTROLS. Where two or more surfaces of revolution such as cylinders, spheres, cones, etc., are generated about a common axis, the amount of permissible deviation from such coaxiality may be expressed by a positional tolerance or a concentricity tolerance. See 5-6.7 for an application of runout tolerance. Selection of the proper control depends on the nature of the functional requirement as follows:

(a) *Positional Tolerance.* Where it is permissible to control a coaxial relationship between two or more features on an MMC basis, positional tolerancing is recommended. See Figure 133. The datum feature may be specified either on an MMC or an RFS basis, depending upon the design requirements. Where positional tolerancing is applied for control of coaxiality, the concepts of 5-5.5.1.2 apply.

FIG. 126 DIFFERENT POSITIONAL TOLERANCE AT ONE END OF HOLE THAN AT OTHER END

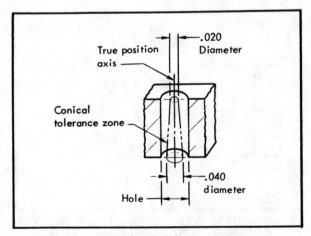

FIG. 127 TOLERANCE ZONE FOR FIGURE 126

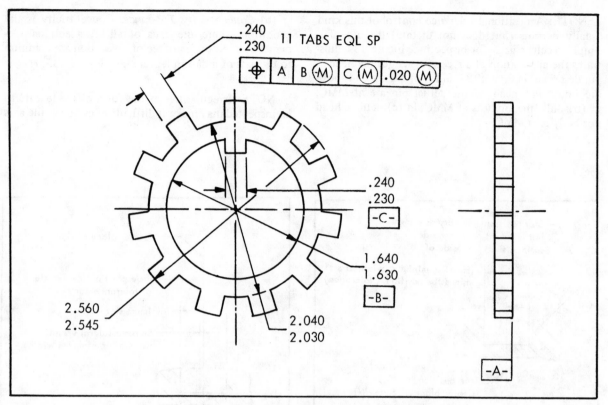

FIG. 128 POSITIONAL TOLERANCING OF TABS

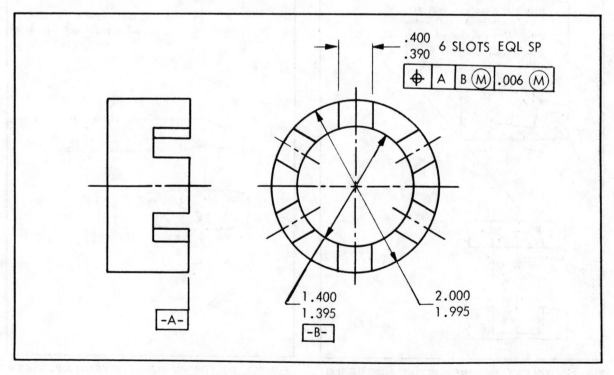

FIG. 129 POSITIONAL TOLERANCING OF SLOTS

NOTE: A positional tolerance control of this kind usually permits (but does not dictate) the use of a simple receiver gage for inspection. Figure 134 illustrates the application of a receiver gage for the headed pin described in Figure 133. In (a) of Figure 134, both the body and head of the actual pin are at MMC; in (b) only the body is at MMC; in (c) neither head nor body is at MMC.

(b) *Concentricity Tolerance.* Concentricity is the condition where the axes of all cross-sectional elements of a feature's surface of revolution are common to the axis of a datum feature. See Figure 135.

NOTE: Irregularities in the form of the feature to be inspected may make it difficult to establish the axis

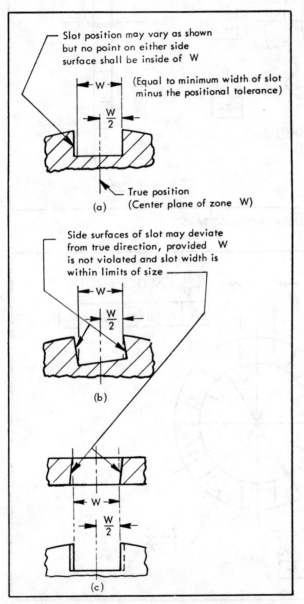

FIG. 130   TOLERANCE ZONE FOR SURFACES OF SLOT
AT MMC

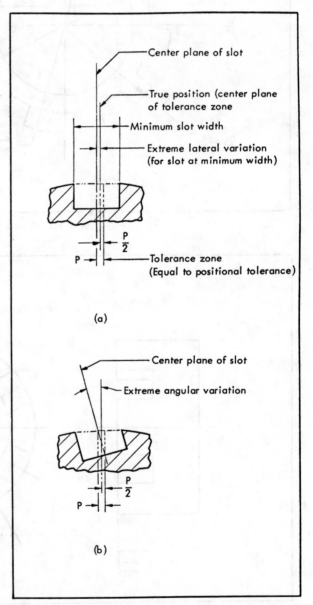

FIG. 131   TOLERANCE ZONE FOR CENTER PLANE OF
SLOT AT MMC

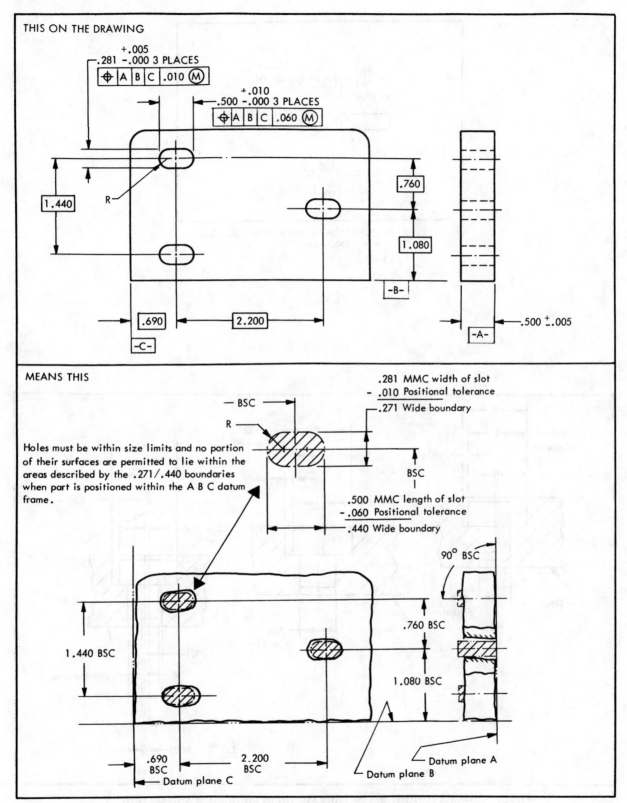

THIS ON THE DRAWING

MEANS THIS

Holes must be within size limits and no portion of their surfaces are permitted to lie within the areas described by the .271/.440 boundaries when part is positioned within the A B C datum frame.

.281 MMC width of slot
− .010 Positional tolerance
.271 Wide boundary

.500 MMC length of slot
− .060 Positional tolerance
.440 Wide boundary

FIG. 132 POSITIONAL TOLERANCING OF ELONGATED HOLES, BOUNDARY CONCEPT

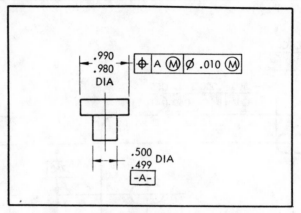

FIG. 133   POSITIONAL TOLERANCING FOR
COAXIALITY

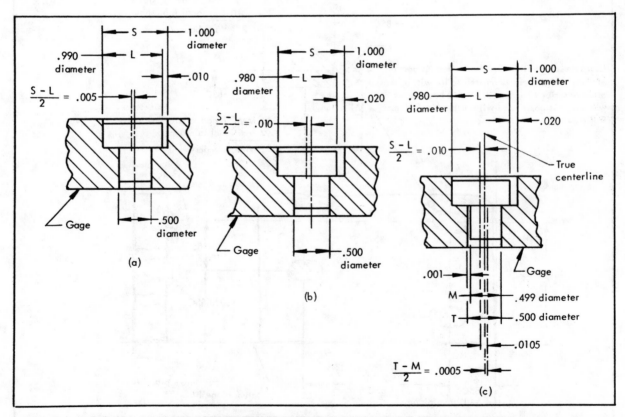

FIG. 134   VARIOUS CONDITIONS OF PART SHOWN IN FIGURE 133 AND MATING PART

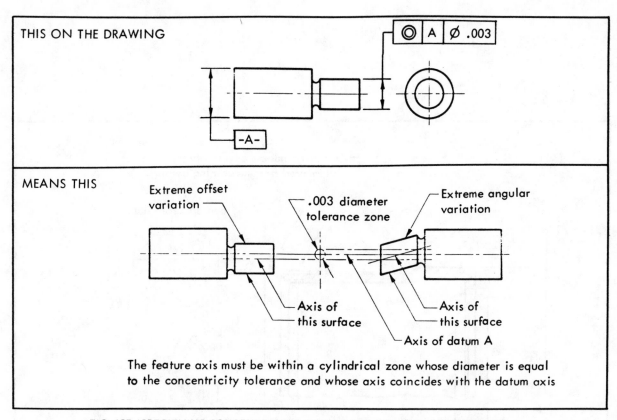

FIG. 135   SPECIFYING CONCENTRICITY, DATUM AND RELATED FEATURE RFS

of the feature. For example, a nominally cylindrical surface may be bowed or out of round in addition to being offset from its datum feature; in such cases, finding the axis of the feature may entail a time-consuming analysis of the surface. Therefore, unless there is a definite need for the control of axes (as in the case shown in Figure 136), it is recommended that control be specified in terms of a runout tolerance (see 5-6.7) or a positional tolerance as in (a) above.

5-5.12.1 Alignment of Coaxial Holes. A positional tolerance is used to control the alignment of two or more holes shown on a common axis. It is used where a tolerance of location alone does not provide the necessary control of alignment of these holes, and a separate requirement must be specified. Figure 137 shows an example of four coaxial holes of the same

size. Where holes are of different specified sizes and the same requirements apply to all holes, a single feature control symbol, supplemented by a notation such as ALL COAXIAL HOLES, is used.

5-5.13 SYMMETRY. Symmetry is a condition in which a feature (or features) is symmetrically disposed about the center plane of a datum feature.

5-5.13.1 Symmetry Tolerance. Where it is required that a feature be located symmetrically with respect to a datum feature, positional tolerancing is recommended. This permits the tolerance to be expressed on an MMC basis (as shown in Figure 138) or on an RFS basis (as shown in Figure 139). Where desired to use a symmetry tolerance and the symmetry symbol, the method in Figure 140 may be followed.

61

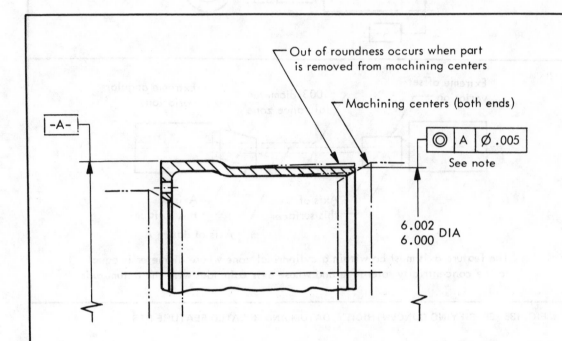

Note: A concentricity tolerance (rather than a runout tolerance, or a positional tolerance) has been applied to the item shown above because of the following supposed conditions:

1.   A precise degree of coaxiality is required when part is assembled with mating parts.

2.   The toleranced diameter, when removed from supporting tooling (machining centers) is likely to go out-of-round to the full amount permitted by the limits of size. This would preclude verification by either a runout inspection or a positional inspection and would require careful analysis of the surface; see 5-5.12(b).

FIG. 136 EXAMPLE WHERE CONCENTRICITY TOLERANCE IS REQUIRED

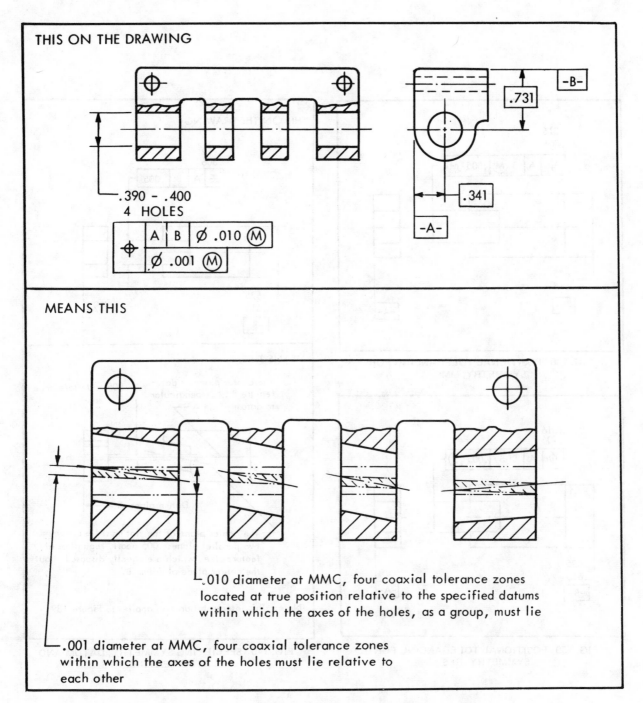

FIG. 137 POSITIONAL TOLERANCING FOR COAXIAL HOLES

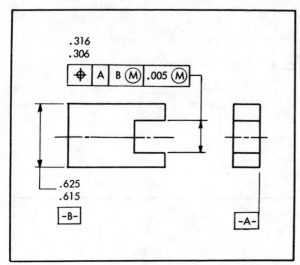

FIG. 138   POSITIONAL TOLERANCING FOR
SYMMETRY, MMC

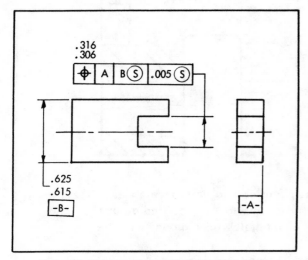

FIG. 139   POSITIONAL TOLERANCING FOR
SYMMETRY, RFS

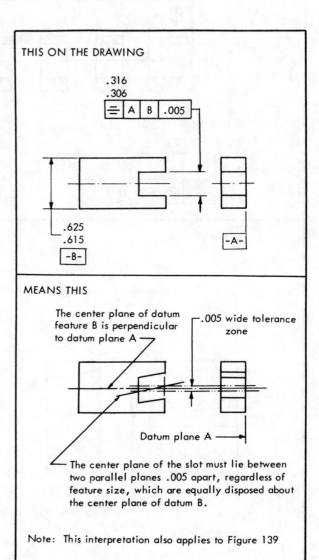

FIG. 140   SPECIFYING SYMMETRY, DATUM AND
RELATED FEATURE RFS

64

5-6   TOLERANCES OF FORM AND RUNOUT

5-6.1 GENERAL. This subsection establishes the principles and methods of dimensioning and tolerancing for control of form and runout of various geometrical shapes, and free-state variations.

5-6.2 FORM CONTROL. Form tolerances control straightness, flatness, roundness, cylindricity, profile of a surface or line, angularity, parallelism and perpendicularity. See Figure 73. Since surface texture and tolerances of size and location control form to a certain degree, the extent of this control should be clearly understood before specifying form tolerances.

5-6.3 SPECIFYING FORM TOLERANCES. Form tolerances are specified for features critical to function and interchangeability where the tolerances of size and location do not provide the required control. The symbolic method of specifying form tolerances is illustrated in this subsection. Equivalent drawing notes were formerly used on an optional basis to express these requirements.

5-6.3.1 Form Tolerance Zones. A form tolerance specifies a tolerance zone within which the considered feature, its axis, or center plane must be contained.

5-6.3.1.1 Where the form tolerance value represents the diameter of a cylindrical zone, it is preceded by the diameter symbol. Otherwise, no symbol is required and the form tolerance value represents a linear distance between two geometric boundaries.

5-6.3.1.2 Certain designs require control of form over a limited area or length of the surface, rather than control of the total surface. In these instances, the area, or length, and the location are identified on a drawing with appropriate dimensioning.

5-6.3.1.3 It may be desirable to control the form of a surface on a unit basis in terms of variation per unit. See Figure 144.

5-6.3.2 Zero Form Tolerance at MMC. Where variations of form for related features (such as perpendicularity) are to be contained within the size limits

of a feature, the feature control symbol contains zeros for the form tolerance, modified by the symbol for MMC. If the feature is finished on its MMC limit of size, it must be perfect in form with respect to the datum. A tolerance of form can exist only as the feature departs from MMC, and the allowable form tolerance is equal to the amount of such departure. See Figures 172 and 173.

5-6.4 FORM TOLERANCES FOR SINGLE FEATURES. Straightness, flatness, roundness, cylindricity, and in some instances, profile are form tolerances applicable to single features.

5-6.4.1 Straightness Tolerance. Straightness is a condition where an element of a surface or an axis is a straight line. A straightness tolerance specifies a tolerance zone within which an axis or all points of the considered element must lie. A straightness tolerance is applied in the view where the elements to be controlled are represented by a straight line.

5-6.4.1.1 Figure 141 shows an example of a cylindrical feature where all circular elements of the surface are to be within the specified size tolerance. Each longitudinal element of the surface must lie in a tolerance zone defined by two parallel lines spaced apart by the amount of the prescribed straightness tolerance, where the two lines and the nominal axis of the feature share a common plane. The symbol is directed to the feature surface or extension line but not to the size dimension. The straightness tolerance must be less than the size tolerance. The principles of 5-2.11.1 (perfect form at MMC) apply.

5-6.4.1.2 Figures 142 and 143 show examples of cylindrical features where all circular elements of the surface are to be within specified size tolerance and the boundary of perfect form at MMC can be violated. This violation is permissible when the feature control symbol is located with the size dimension or attached to the dimension line. In this instance, a diameter symbol precedes the tolerance value and the tolerance is applied on either an RFS or MMC basis. Where either RFS or MMC basis applies, the collective effect of size and form variation can produce a virtual condition equal to the MMC size plus the straightness tolerance. When applied on an RFS basis, as in Figure 142, the maximum straightness tolerance is the specified tolerance. When applied on an MMC basis, as in Figure 143, the maximum straightness tolerance is the specified tolerance plus the amount the feature departs from its MMC size. As established from the surface elements, the derived axis or center-line of the actual features must lie within a cylindrical tolerance zone of the amount prescribed above.

5-6.4.1.3 The application of straightness on a unit basis is a means of preventing an abrupt surface variation within a relatively short length of the feature. See Figure 144. Caution should be exercised when using unit control because of relatively large theo-retical variations that may result if left unrestricted. If the unit variation appears as a "bow" in the toleranced feature, and the "bow" is allowed to continue for several units, the ultimate tolerance variation may pro-duce an unsatisfactory part. Figure 145 illustrates the possible condition if straightness per unit length given in Figure 144 is used alone. That is, straightness for the total length is not specified.

5-6.4.1.4 Figure 146 illustrates the use of straightness tolerance on a flat surface. Straightness may be applied to control line elements in a single direction on a flat surface; it may also be applied in two direc-tions as shown. As an extension of the principles of 5-6.4.1.2, straightness may be applied on an RFS or MMC basis to non-cylindrical features of size. In this instance, the derived center plane, as established from the surface elements of the actual feature, must lie in a tolerance zone between two parallel planes separated

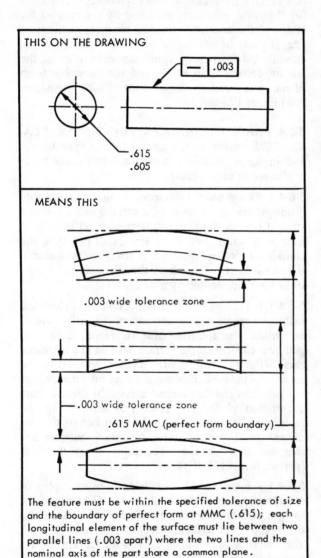

THIS ON THE DRAWING

MEANS THIS

.003 wide tolerance zone

.003 wide tolerance zone

.615 MMC (perfect form boundary)

The feature must be within the specified tolerance of size and the boundary of perfect form at MMC (.615); each longitudinal element of the surface must lie between two parallel lines (.003 apart) where the two lines and the nominal axis of the part share a common plane.

FIG. 141 SPECIFYING STRAIGHTNESS OF SURFACE ELEMENTS

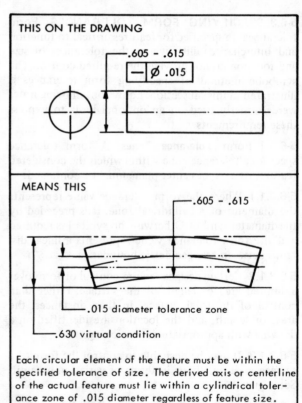

THIS ON THE DRAWING

.605 – .615

MEANS THIS

.605 – .615

.015 diameter tolerance zone

.630 virtual condition

Each circular element of the feature must be within the specified tolerance of size. The derived axis or centerline of the actual feature must lie within a cylindrical toler-ance zone of .015 diameter regardless of feature size.

FIG. 142 SPECIFYING STRAIGHTNESS RFS

by the amount of the tolerance. Symbol placement and arrangement as described in 5-6.4.1.2 applies except the diameter symbol is not used as the tolerance zone is non-cylindrical.

5-6.4.2 Flatness Tolerance. Flatness is the condition of a surface having all elements in one plane. A flatness tolerance specifies a tolerance zone defined by two parallel planes within which the surface must lie. When a flatness tolerance is specified, the symbol is attached to a leader directed to the surface or to an extension line of the surface. The symbol is placed in a view where the surface elements to be controlled are represented by a line. See Figure 147.

5-6.4.2.1 Flatness may be applied on a unit basis as a means of preventing an abrupt surface variation with-in a relatively short length of the feature. The unit variation is used either in combination with a specified total variation, or alone. Caution should be exercised when using unit control alone for the reasons given in 5-6.4.1.3. Since flatness involves surface area, the drawing should identify the unit area applicable.

5-6.4.3 Roundness (Circularity) Tolerance. Roundness is a condition of a surface of revolution where:

(a) For a cylinder or cone, all points of the surface intersected by any plane perpendicular to a common axis are equidistant from that axis.

(b) For a sphere, all points of the surface intersected by any plane passing through a common center are equidistant from that center.

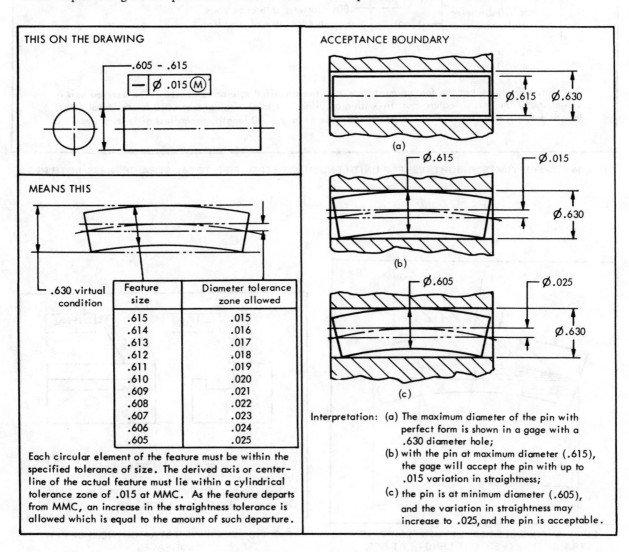

FIG. 143 SPECIFYING STRAIGHTNESS AT MMC

67

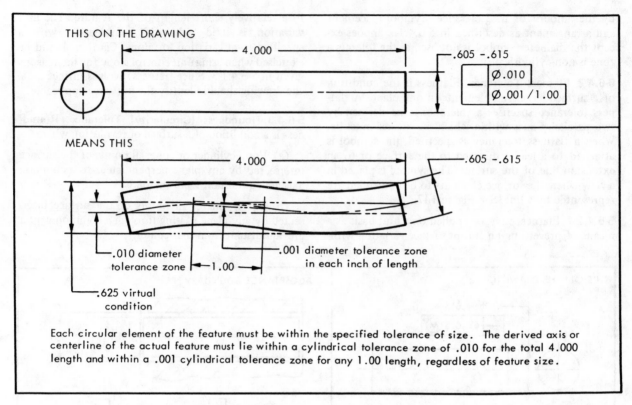

THIS ON THE DRAWING

MEANS THIS

.010 diameter tolerance zone

.001 diameter tolerance zone in each inch of length

.625 virtual condition

Each circular element of the feature must be within the specified tolerance of size. The derived axis or centerline of the actual feature must lie within a cylindrical tolerance zone of .010 for the total 4.000 length and within a .001 cylindrical tolerance zone for any 1.00 length, regardless of feature size.

FIG. 144 SPECIFYING STRAIGHTNESS PER UNIT LENGTH WITH SPECIFIED TOTAL STRAIGHTNESS, BOTH RFS

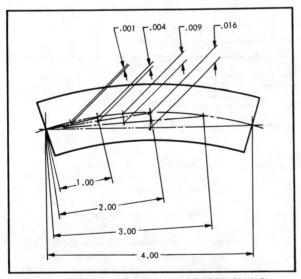

FIG. 145 POSSIBLE RESULTS OF SPECIFYING STRAIGHTNESS PER UNIT LENGTH AT RFS, WITH NO SPECIFIED TOTAL

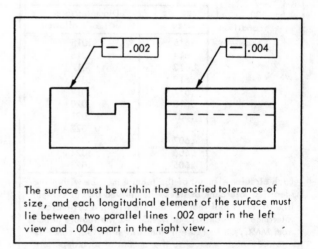

The surface must be within the specified tolerance of size, and each longitudinal element of the surface must lie between two parallel lines .002 apart in the left view and .004 apart in the right view.

FIG. 146 SPECIFYING STRAIGHTNESS OF FLAT SURFACES

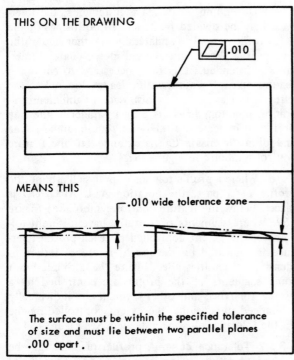

THIS ON THE DRAWING

MEANS THIS

.010 wide tolerance zone

The surface must be within the specified tolerance
of size and must lie between two parallel planes
.010 apart.

FIG. 147 SPECIFYING FLATNESS

A roundness tolerance specifies a tolerance zone bounded by two concentric circles within which each circular element of the surface must lie, and applies independently at any plane described in (a) and (b) above. See Figures 148 and 149.

5-6.4.4 Cylindricity Tolerance. Cylindricity is a condition of a surface of revolution in which all points of the surface are equidistant from a common axis. A cylindricity tolerance specifies a tolerance zone bounded by two concentric cylinders within which the surface must lie. In the case of cylindricity, unlike that of roundness, the tolerance applies simultaneously to both circular and longitudinal elements of the surface (the entire surface). See Figure 150. The leader from the feature control symbol may be directed to either view.

NOTE: The cylindricity tolerance is a composite control of form which includes roundness, straightness, and taper of a cylindrical feature.

5-6.5 PROFILE TOLERANCE. A profile is the outline of an object in a given plane (two dimensional) figure. Profiles are formed by projecting a three

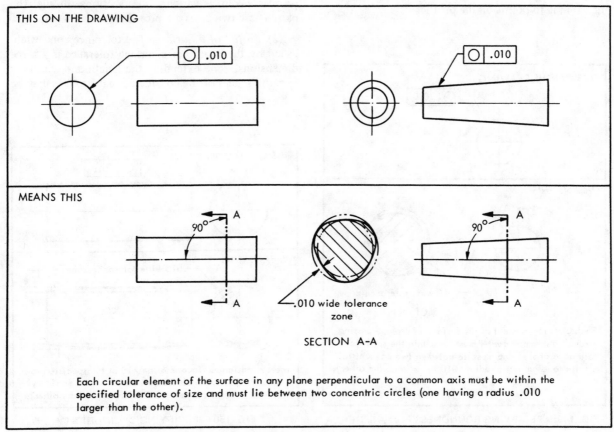

THIS ON THE DRAWING

MEANS THIS

90° A

A

.010 wide tolerance
zone

90° A

A

SECTION A-A

Each circular element of the surface in any plane perpendicular to a common axis must be within the specified tolerance of size and must lie between two concentric circles (one having a radius .010 larger than the other).

FIG. 148 SPECIFYING ROUNDNESS FOR A CYLINDER OR CONE

dimensional figure onto a plane or by taking cross sections through the figure. The elements of a profile are straight lines, arcs and other curved lines. If the drawing specifies individual tolerances for the elements (or points) of a profile, these elements (or points) must be individually verified. Such a procedure may be impractical in certain cases, particularly where accuracy of the entire profile (rather than elements of a profile) is a design requirement. With profile tolerancing, the true profile is defined by basic dimensions (without tolerances) and the tolerance specifies a uniform boundary along this profile within which elements of the surface must lie. Profile tolerancing is used as a form control or a combination of size and form control. Where used as a refinement of size, the profile tolerance must be contained within the size tolerance.

5-6.5.1 Application. Profile tolerances are specified as follows:

(a) An appropriate view or section is drawn showing the desired basic profile in true shape.

(b) The profile is defined by basic dimensions. This dimensioning may be in the form of located radii and angles, or it may consist of coordinate dimensioning to points on the profile.

(c) Depending on design requirements, the tolerance may be divided bilaterally to both sides of the true profile or applied unilaterally to either side of the true profile. Where an equally disposed bilateral tolerance is intended, it is only necessary to show the feature control symbol with a leader directed to the surface. For an unequally disposed or a unilateral tolerance, phantom lines are drawn parallel to the true profile to indicate the tolerance zone boundary. One end of a dimension line is extended to the feature control symbol. See Figure 151.

(d) Where a profile tolerance applies all around the profile of a part, the notation ALL AROUND is placed below the feature control symbol. See Figure 152. Where segments of a profile have different tolerances, the extent controlled by each profile tolerance is indicated by the use of reference letters to identify extremities. See Figure 153. Similarly, if some segments of the profile are controlled by a profile tolerance and other segments by individually toleranced dimensions, the extent controlled by the profile tolerance must be indicated. See Figure 154.

5-6.5.2 Tolerance Zone. A profile tolerance may be applied to either an entire surface or to individual profiles taken at various cross sections through the part. These two cases are provided for as follows:

(a) *Profile of a Surface.* The tolerance zone established by the profile of a surface tolerance is a three dimensional zone extending the length and width (or circumference) of the considered feature. This may be

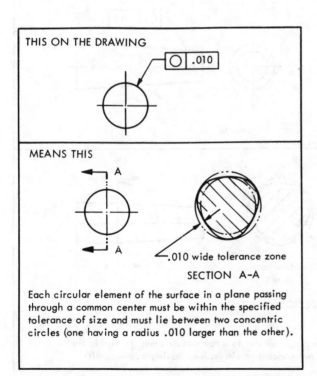

THIS ON THE DRAWING

⌀ .010

MEANS THIS

SECTION A-A

.010 wide tolerance zone

Each circular element of the surface in a plane passing through a common center must be within the specified tolerance of size and must lie between two concentric circles (one having a radius .010 larger than the other).

FIG. 149   SPECIFYING ROUNDNESS FOR A SPHERE

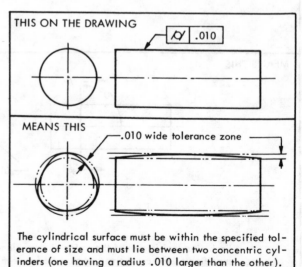

THIS ON THE DRAWING

⌭ .010

MEANS THIS

.010 wide tolerance zone

The cylindrical surface must be within the specified tolerance of size and must lie between two concentric cylinders (one having a radius .010 larger than the other).

FIG. 150   SPECIFYING CYLINDRICITY

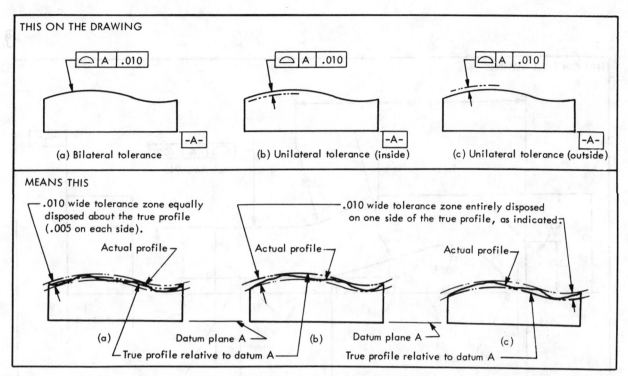

FIG. 151 APPLICATION OF PROFILE OF A SURFACE TOLERANCE TO A BASIC CONTOUR

applied to parts having a constant cross section throughout the thickness as in Figure 152 or to parts having a surface of revolution.

(b) *Profile of a Line.* The tolerance zone established by the profile of a line tolerance is a two dimensional zone extending along the length of the considered feature. This applies to the profiles of parts having a varying cross section, such as the wing of an aircraft, or to random cross sections of parts where it is not desired to control the entire surface of the feature as a single entity as in Figures 157 and 158.

5-6.5.3 Explanation. The tolerance value represents the distance between two boundaries either disposed about the true profile or entirely disposed on one side of the true profile. Profile tolerances apply normal (perpendicular) to the true profile at all points along the profile. The boundaries of the tolerance zone follow the geometric shape of the true profile. Where a profile tolerance encompasses a sharp corner, it is understood to extend to the intersection of the outside boundary lines. See Figure 155. In such cases, the corner radius of the part may also be specified. See Figure 152. The actual surface or line element must be within the specified tolerance zone and all variations from the true profile must blend.

5-6.5.4 Application of Datums. In most cases, profile of a surface tolerancing requires reference to datums in order to provide proper orientation of the profile. With profile of a line tolerancing, datums may be used under some circumstances but would not be used when the only requirement is the profile shape taken cross section by cross section, for example, the shape of a continuous extrusion.

5-6.5.5 Combined Controls. Profile tolerancing may be combined with other types of form tolerancing. Figure 156 illustrates a surface which has a profile tolerance refined by a parallelism tolerance. Each line element must not only lie within the profile tolerance but must also be parallel to the datum within the tolerance specified. Figure 157 illustrates a surface which has a profile tolerance refined by a runout tolerance. Any longitudinal element of the surface must lie within the profile tolerance and any circular element must also be within the specified runout tolerance. Figure 158 shows a part with a profile of a line tolerance in which size is controlled by a separate tolerance. In this case, the profile as an entity may fall anywhere within the size tolerance zone, but the shape of the profile must also fall within the profile tolerance zone.

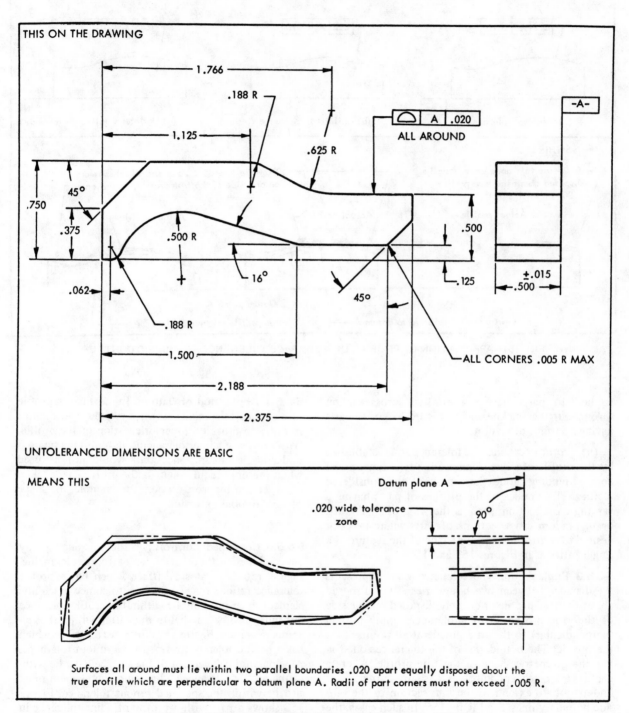

THIS ON THE DRAWING

ALL AROUND

-A-

UNTOLERANCED DIMENSIONS ARE BASIC

MEANS THIS

Datum plane A

.020 wide tolerance zone

90°

Surfaces all around must lie within two parallel boundaries .020 apart equally disposed about the true profile which are perpendicular to datum plane A. Radii of part corners must not exceed .005 R.

FIG. 152   SPECIFYING PROFILE OF A SURFACE ALL AROUND

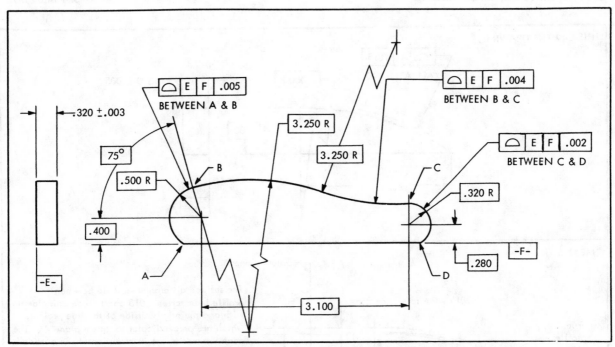

FIG. 153   SPECIFYING DIFFERENT TOLERANCES ON SEGMENTS OF A PROFILE

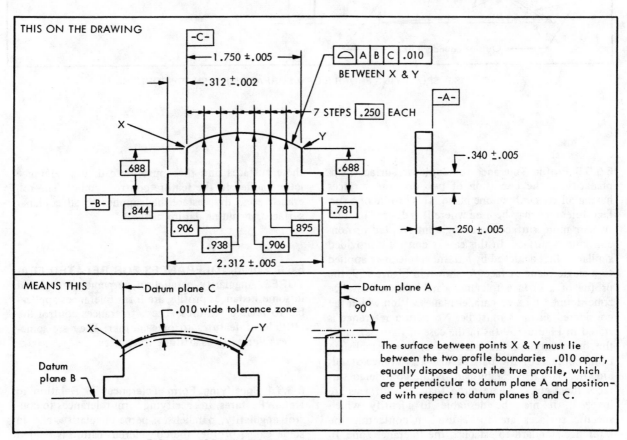

FIG. 154   SPECIFYING PROFILE OF A SURFACE BETWEEN POINTS

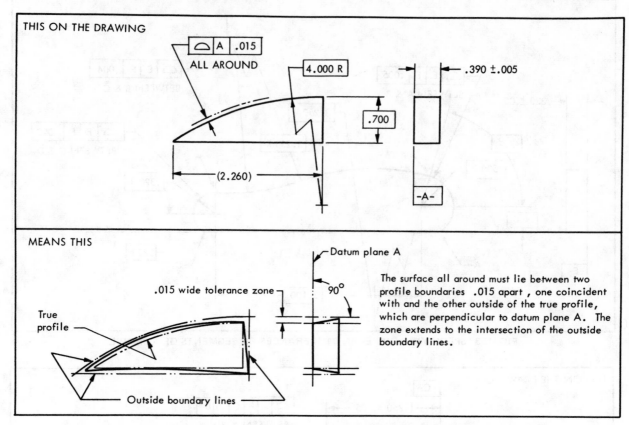

FIG. 155 SPECIFYING PROFILE OF A SURFACE WITH SHARP CORNERS

5-6.5.6 Profile Tolerance for Coplanar Surfaces. Coplanarity is the condition of two or more surfaces having all elements in one plane. This profile of a surface tolerance may be used where it is desired to treat two or more surfaces as a single interrupted or noncontinuous surface. In this case, a control is provided similar to that achieved by a flatness tolerance applied to a single plane surface. As shown in Figure 159, the profile of a surface tolerance establishes a tolerance zone defined by two parallel planes within which the considered surfaces must lie. No datum reference is stated in Figure 159 (as in the case of flatness), since the orientation of the tolerance zone is established from contact of the part against a reference standard; the datum is established by the considered surfaces themselves. Where more than two surfaces are involved, it may be desirable to identify which specific surfaces are to be used in contacting the reference standard to establish the tolerance zone. In this case, datum identifying symbols are applied to

these surfaces and the appropriate datum reference letters added to the feature control symbol. The tolerance zone thus established applies to all coplanar surfaces, including datum surfaces.

5-6.6 FORM TOLERANCES FOR RELATED FEATURES. Angularity, parallelism, perpendicularity, and in some instances profile, are form tolerances applicable to related features. These tolerances control the attitude of features to one another. They are sometimes referred to as attitude tolerances.

5-6.6.1 Specifying Form Tolerances in Relation to Datum Features. In specifying form tolerances to control angularity, parallelism, perpendicularity, and in some cases, profile, the considered feature is related to one or more datum features. Relation to more than

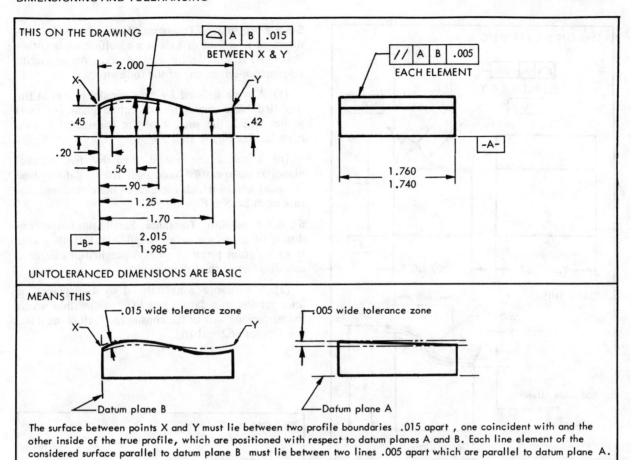

FIG. 156 SPECIFYING COMBINED PROFILE AND PARALLELISM TOLERANCES

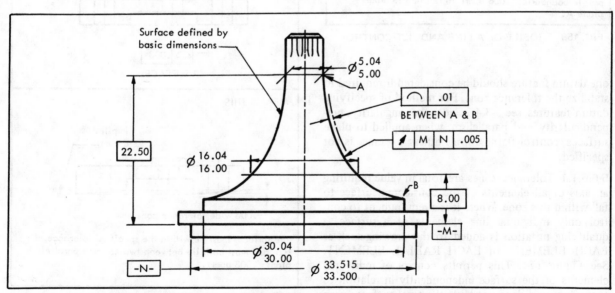

FIG. 157 SPECIFYING COMBINED PROFILE AND RUNOUT TOLERANCES

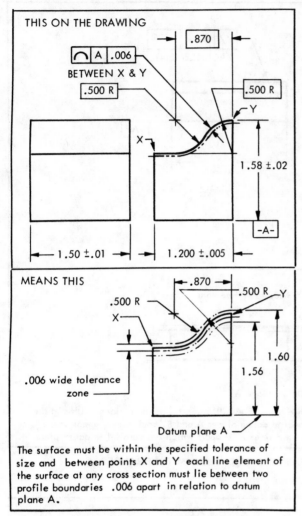

FIG. 158   PROFILE OF A LINE AND SIZE CONTROL

one datum feature should be considered if required to stabilize the tolerance zone. For method of specifying datum features, see 5-3.4.2. Note that angularity, perpendicularity and parallelism, when applied to plane surfaces, control flatness if a flatness tolerance is not specified.

5-6.6.1.1 Tolerance zones are total in value requiring an axis or all elements of the considered surface to fall within this zone. Where it is a requirement to control only individual line elements of a surface, a qualifying notation is added to the drawing such as EACH ELEMENT or EACH RADIAL ELEMENT. See Figure 174. This permits control of individual elements of the surface independently in relation to the datum and does not limit the total surface to an encompassing zone.

5-6.6.2 Angularity Tolerance. Angularity is the condition of a surface or axis at a specified angle (other than 90°) from a datum plane or axis. An angularity tolerance specifies one of the following:

(a) A zone defined by two parallel planes at the specified basic angle from a datum plane (or axis) within which the surface of the considered feature must lie. See Figure 160.

(b) A tolerance zone defined by two parallel planes at the specified basic angle from a datum plane (or axis) within which the axis of the considered feature must lie. See Figure 161.

5-6.6.3 Parallelism Tolerance. Parallelism is the condition of a surface or axis equidistant at all points from a datum plane or axis. A parallelism tolerance specifies one of the following:

(a) A tolerance zone defined by two planes or lines parallel to a datum plane (or axis) within which the surface or axis of the considered feature must lie. See Figures 162 and 163.

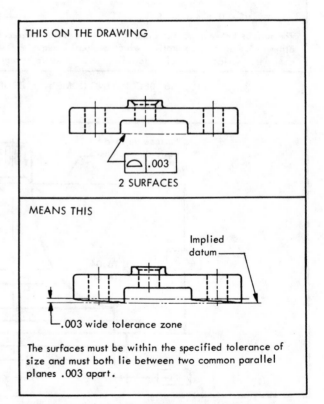

FIG. 159   SPECIFYING PROFILE OF A SURFACE FOR COPLANAR SURFACES

(b) A cylindrical tolerance zone parallel to a datum axis within which the axis of the considered feature must lie. See Figures 164 and 165.

5-6.6.4 Perpendicularity Tolerance. Perpendicularity is the condition of a surface, median plane, or axis at a right angle to a datum plane or axis. A perpendicularity tolerance specifies one of the following:

(a) A tolerance zone defined by two parallel planes perpendicular to a datum plane or axis within which the surface or median plane of the considered feature must lie. See Figures 166 and 167.

(b) A tolerance zone defined by two parallel planes perpendicular to a datum axis within which the axis of the considered feature must lie. See Figure 168.

(c) A cylindrical tolerance zone perpendicular to a datum plane within which the axis of the considered feature must lie. See Figures 169 through 173.

(d) A tolerance zone defined by two parallel lines perpendicular to a datum plane or axis within which an element of the surface must lie. See Figure 174.

5-6.7 RUNOUT TOLERANCE. Runout is a composite tolerance used to control the functional relationship of one or more features of a part to a datum axis. The types of features controlled by runout tolerances include those surfaces constructed around a datum axis and those constructed at right angles to a datum axis. See Figure 175.

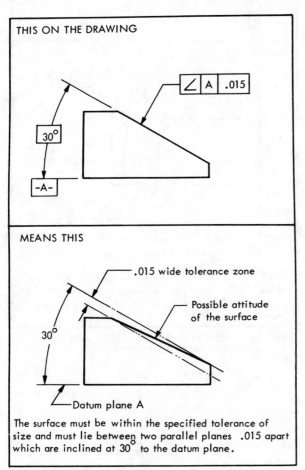

FIG. 160 SPECIFYING ANGULARITY FOR A PLANE SURFACE

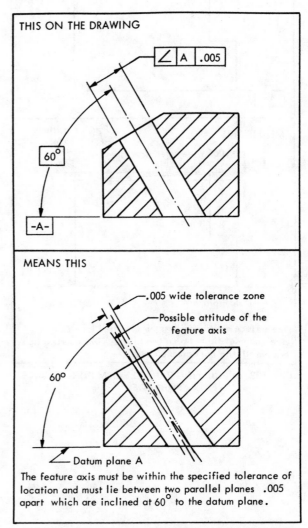

FIG. 161 SPECIFYING ANGULARITY FOR AN AXIS (FEATURE RFS)

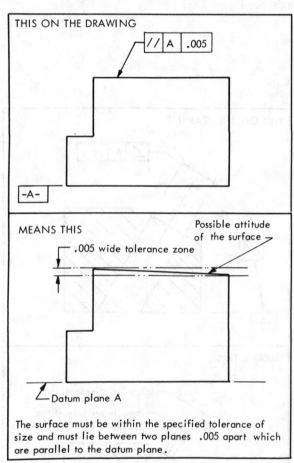

FIG. 162 SPECIFYING PARALLELISM FOR A PLANE
SURFACE

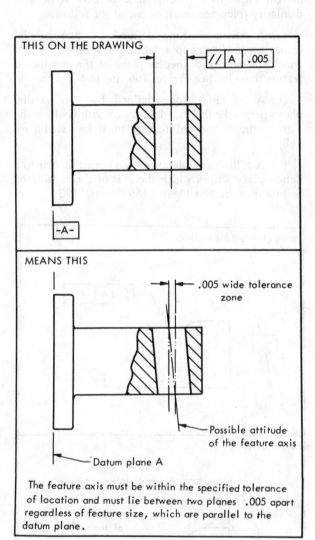

FIG. 163 SPECIFYING PARALLELISM FOR AN AXIS
(FEATURE RFS)

THIS ON THE DRAWING

// | A | ⌀ .005

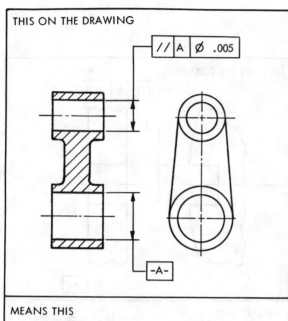

-A-

THIS ON THE DRAWING

⌀ .264 – .267

// | A | ⌀ .002 Ⓜ

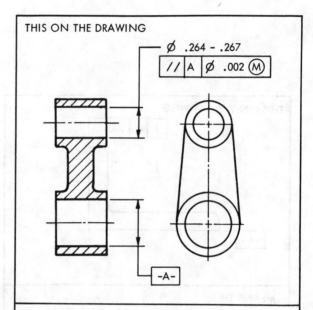

-A-

MEANS THIS

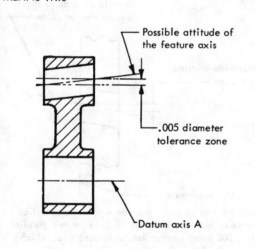

Possible attitude of
the feature axis

.005 diameter
tolerance zone

Datum axis A

The feature axis must be within the specified tolerance
of location and must lie within a cylindrical zone .005
diameter , regardless of feature size, which is parallel
to the datum axis.

FIG. 164 SPECIFYING PARALLELISM FOR AN AXIS
(BOTH FEATURE AND DATUM FEATURE RFS)

MEANS THIS

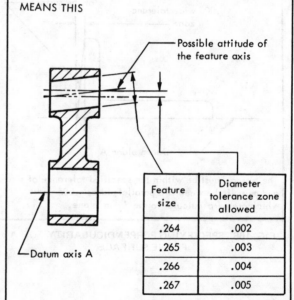

Possible attitude of
the feature axis

Datum axis A

Feature size	Diameter tolerance zone allowed
.264	.002
.265	.003
.266	.004
.267	.005

The feature axis must be within the specified tolerance of
location. Where the feature is at maximum material condition (.264), the maximum parallelism tolerance is .002
diameter. Where the feature departs from its MMC size,
an increase in the parallelism tolerance is allowed which
is equal to the amount of such departure.

FIG. 165 SPECIFYING PARALLELISM FOR AN AXIS
(FEATURE AT MMC AND DATUM FEATURE RFS)

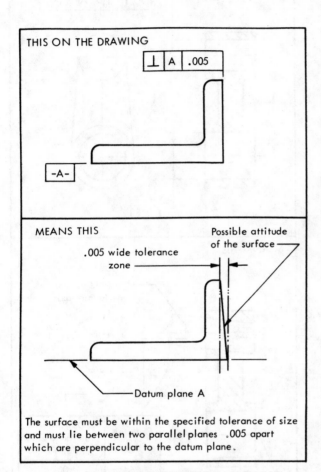

FIG. 166 SPECIFYING PERPENDICULARITY FOR A
PLANE SURFACE

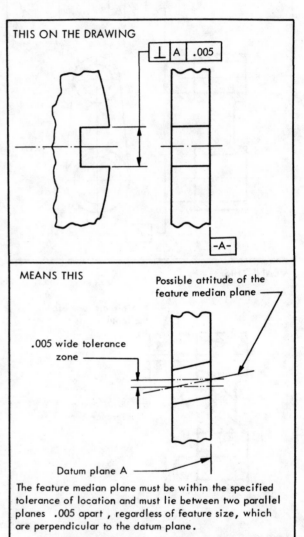

FIG. 167 SPECIFYING PERPENDICULARITY FOR A
MEDIAN PLANE (FEATURE RFS)

80

THIS ON THE DRAWING

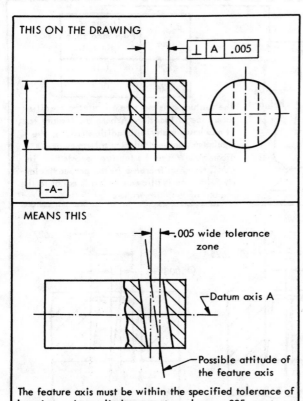

MEANS THIS

.005 wide tolerance
zone

Datum axis A

Possible attitude of
the feature axis

The feature axis must be within the specified tolerance of
location and must lie between two planes .005 apart,
regardless of feature size, which are perpendicular to the
datum axis.

Note: This tolerance applies only to the view on which
it is specified, not the end view.

FIG. 168 SPECIFYING PERPENDICULARITY FOR AN
AXIS (BOTH FEATURE AND DATUM FEATURE RFS)

THIS ON THE DRAWING

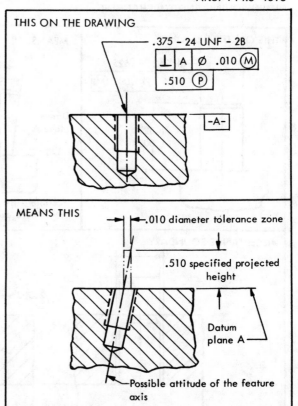

MEANS THIS

.010 diameter tolerance zone

.510 specified projected
height

Datum
plane A

Possible attitude of the feature
axis

The feature pitch diameter axis must be within the specified tolerance of location and must lie within a cylindrical zone .010 diameter at MMC which is perpendicular
to and projects from the datum plane for the .510 specified height.

Note: Threaded holes are considered to be located and
gaged from their pitch diameters, normally at MMC.
Consideration must be given to the additive tolerance as a
result of the departure from MMC of the thread pitch diameter. The centering effect of the fastener at assembly,
however, may reduce or negate such added tolerance.

(Also see 5-5.7)

FIG. 169 SPECIFYING PERPENDICULARITY FOR AN
AXIS AT A PROJECTED HEIGHT
(THREADED HOLE OR INSERT AT MMC)

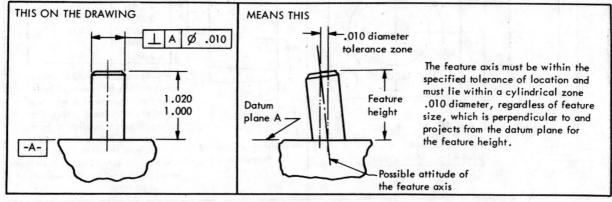

The feature axis must be within the
specified tolerance of location and
must lie within a cylindrical zone
.010 diameter, regardless of feature
size, which is perpendicular to and
projects from the datum plane for
the feature height.

FIG. 170 SPECIFYING PERPENDICULARITY FOR AN AXIS (PIN OR BOSS RFS)

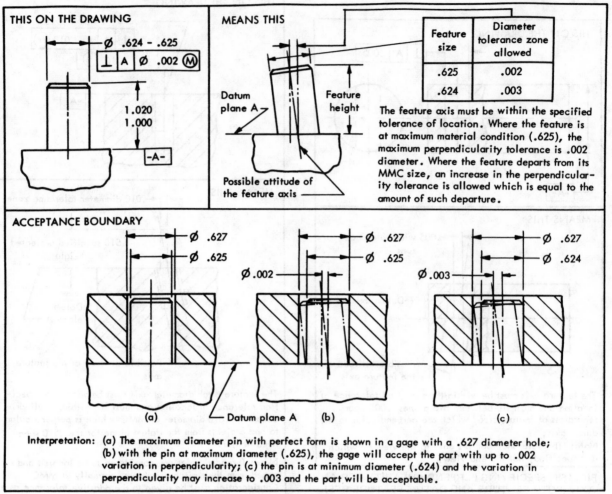

FIG. 171 SPECIFYING PERPENDICULARITY FOR AN AXIS SHOWING ACCEPTANCE
BOUNDARY (PIN OR BOSS AT MMC)

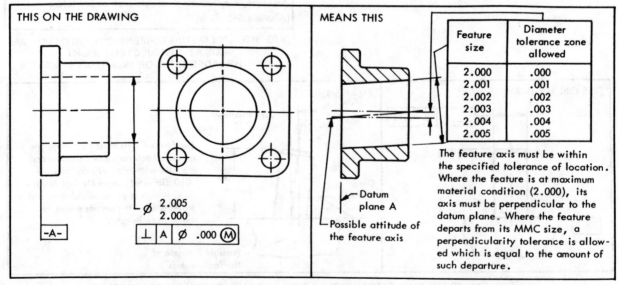

FIG. 172 SPECIFYING PERPENDICULARITY FOR AN AXIS (ZERO TOLERANCE AT MMC)

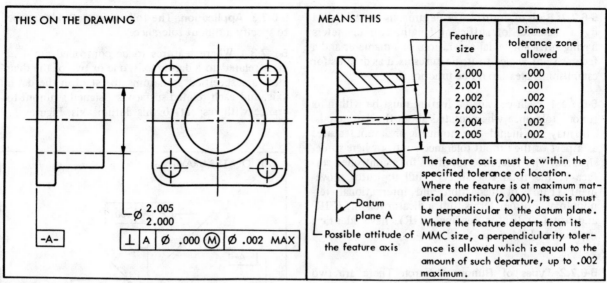

Feature size	Diameter tolerance zone allowed
2.000	.000
2.001	.001
2.002	.002
2.003	.002
2.004	.002
2.005	.002

The feature axis must be within the specified tolerance of location. Where the feature is at maximum material condition (2.000), its axis must be perpendicular to the datum plane. Where the feature departs from its MMC size, a perpendicularity tolerance is allowed which is equal to the amount of such departure, up to .002 maximum.

FIG. 173 SPECIFYING PERPENDICULARITY FOR AN AXIS (ZERO TOLERANCE AT MMC WITH A MAXIMUM SPECIFIED)

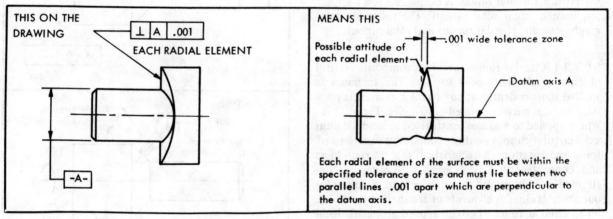

Each radial element of the surface must be within the specified tolerance of size and must lie between two parallel lines .001 apart which are perpendicular to the datum axis.

FIG. 174 SPECIFYING PERPENDICULARITY FOR A RADIAL ELEMENT

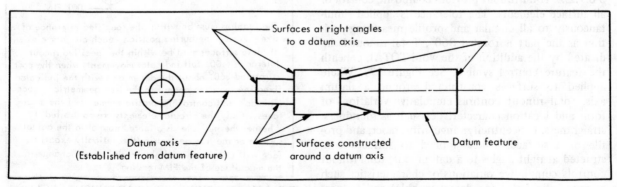

FIG. 175 FEATURES APPLICABLE TO RUNOUT TOLERANCING

5-6.7.1 Basis of Control. The datum axis is established by a diameter of sufficient length, two diameters having sufficient axial separation, or a diameter and a face at right angles to it. Features used as datums for establishing axes should be functional.

5-6.7.1.1 Each considered feature must be within its runout tolerance when rotated about the datum axis. This may also include the mounting or datum features as a part of the runout tolerance control where so designated. The tolerance specified for a controlled surface is the total tolerance or full indicator movement (FIM) in inspection and international terminology. Former terms, full indicator reading (FIR) and total indicator reading (TIR), have identical meaning to FIM.

5-6.7.2 Types of Runout Control. There are two types of runout control, circular runout and total runout. Selection of the type applicable is dependent upon design requirements and manufacturing considerations. Circular runout is normally a less complex requirement than total runout. The following paragraphs describe circular runout and total runout.

5-6.7.2.1 Circular runout provides composite control of circular elements of a surface. The tolerance is applied independently at any circular measuring position as the part is rotated 360°. See Figure 176. Where applied to surfaces constructed around a datum axis, circular runout controls cumulative variations of form and location characteristics, that is, roundness and concentricity. Where applied to surfaces constructed at right angles to a datum axis, circular runout controls circular elements of the surface (wobble). When circular runout is to be applied at specific locations, it is so stated on the drawing.

5-6.7.2.2 Total runout provides composite control of all surface elements. The tolerance is applied simultaneously to all circular and profile measuring position as the part is rotated 360°. Total runout is indicated by the addition of the word TOTAL beneath the feature control symbol. See Figure 177. Where applied to surfaces constructed around a datum axis, total runout controls cumulative variations of form and location characteristics such as roundness, straightness, concentricity, angularity, taper, and profile of a surface. Where applied to surfaces constructed at right angles to a datum axis, total runout controls cumulative variations of characteristics such as perpendicularity (to detect wobble) and flatness (to detect concavity or convexity).

5-6.7.3 Application. The following methods are used to specify a runout tolerance.

5-6.7.3.1 Where features to be controlled are diameters related to a datum axis, one or two of the diameters are specified as datums to establish the datum axis, and each related surface is assigned a runout tolerance with respect to this datum axis. Figures 176

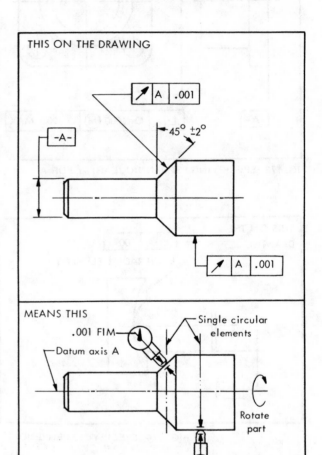

THIS ON THE DRAWING

MEANS THIS

The features must be within the specified tolerance of size. At any measuring position, each circular element of these surfaces must be within the specified runout tolerance (.001 full indicator movement) when the part is rotated 360° about the datum axis with the indicator fixed in a position normal to the true geometric surface. (This does not control the profile elements of these surfaces. Only the circular elements are controlled.) Whether the indicator is oriented normal to the actual surface or the true geometric (theoretically exact) surface will cause only a slight "cosine error" change in the magnitude of the FIM reading.

FIG. 176 SPECIFYING CIRCULAR RUNOUT RELATIVE
TO A DATUM DIAMETER

THIS ON THE DRAWING

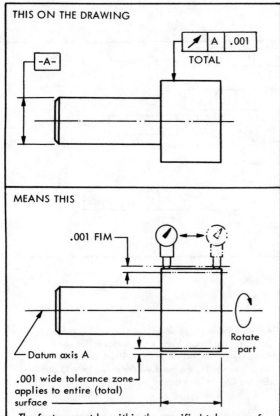

MEANS THIS

The feature must be within the specified tolerance of size. The entire surface must lie within the specified runout tolerance zone (.001 full indicator movement) when the part is rotated 360° about the datum axis with the indicator placed at every location along the surface in a position normal to the true geometric surface, without reset of the indicator. (This controls the cumulative profile and circular elements of the entire surface.) Whether the indicator is oriented normal to the actual surface or the true geometric (theoretically exact) surface will cause only a slight "cosine error" change in the magnitude of the FIM reading.

FIG. 177 SPECIFYING TOTAL RUNOUT RELATIVE TO A DATUM DIAMETER

and 177 illustrate the basic principle of relating features in a runout tolerance to a datum axis as established from a single datum diameter (cylinder) of sufficient length. Figure 176 incorporates the principle of circular runout tolerancing and illustrates the concept of a circular runout tolerance zone applied to features of this kind. Figure 177 incorporates the principle of total runout tolerancing and illustrates the concept of a total runout tolerance zone.

5-6.7.3.2 Figure 178 illustrates application of runout tolerances where two datum diameters collectively establish a single datum axis to which the features are related.

5-6.7.3.3 Where features to be controlled are related to a diameter and a face at right angles to it, each related surface is assigned a runout tolerance with respect to these two datums. The datums are usually specified separately to indicate datum precedence. See Figure 179. This figure incorporates the principles of both circular and total runout tolerancing.

5-6.7.3.4 It may be necessary to control individual datum surface variations with respect to flatness, roundness, parallelism, straightness or cylindricity. Where such control is required, the appropriate form tolerance is specified. See Figures 180 and 181 for examples applying cylindricity and flatness to the datums.

5-6.7.3.5 Where datum features are required by function to be included in the runout control, runout tolerances must be specified for these features. See Figure 180. A runout tolerance specified for the datum features has no effect on the considered features.

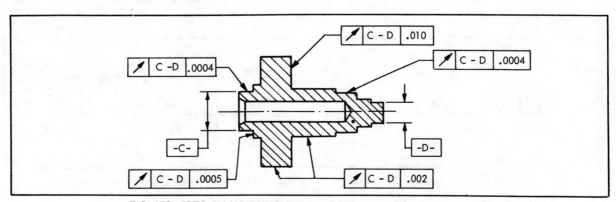

FIG. 178 SPECIFYING RUNOUT RELATIVE TO TWO DATUM DIAMETERS

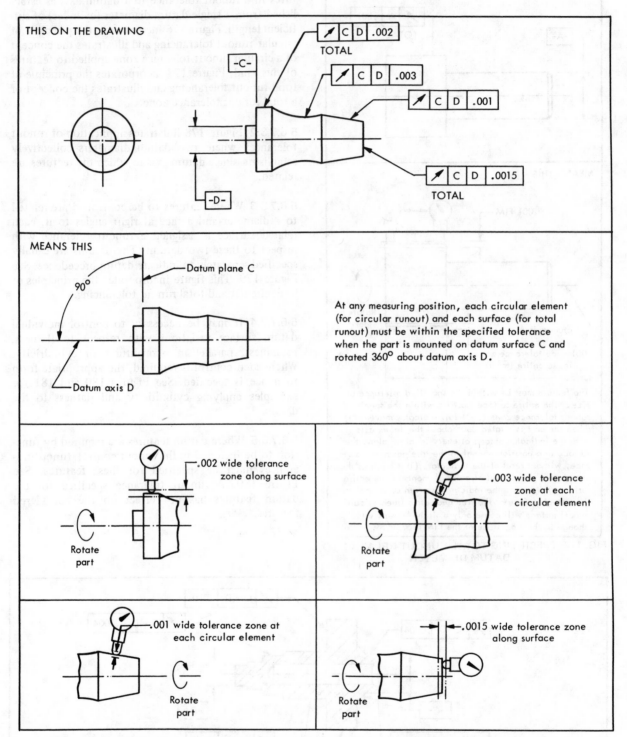

FIG. 179 SPECIFYING RUNOUT RELATIVE TO A DATUM SURFACE AND A DIAMETER

5-6.7.3.6 Features having a specific relationship to each other rather than to a common datum axis are indicated by appropriate datum references within the feature control symbol as shown in Figure 180 where runout tolerance of the hole is related to datum G rather than the axis C-D.

5-6.7.4 Surface Relationship. Any two surfaces on a datum axis individually within their runout tolerances are collectively related to each other within the sum of the indicator readings.

5-6.7.5 Specification. Multiple leaders may be used to direct a feature control symbol to two or more sur-

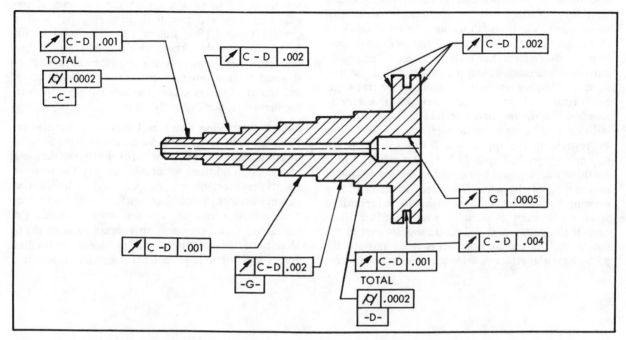

FIG. 180 SPECIFYING RUNOUT RELATIVE TO TWO DATUM DIAMETERS WITH OTHER
FORM CONTROL SPECIFIED

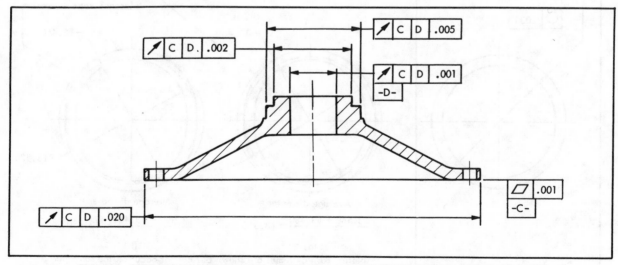

FIG. 181 SPECIFYING RUNOUT RELATIVE TO A DATUM SURFACE AND DIAMETER WITH
OTHER FORM CONTROL SPECIFIED

faces having a common runout tolerance. Surfaces may be specified individually or in groups without affecting the runout tolerance. See Figure 180.

5-6.8 FREE STATE VARIATION. Free state variation is a term used to describe distortion of a part after removal of forces applied during manufacture. This distortion is due principally to the weight and flexibility of the part and the release of internal stresses resulting from fabrication. A part of this kind (for example, a part with a very thin wall in proportion to its diameter) is referred to as a "nonrigid" part. In some cases, it may be required that the part meet its tolerance requirements while in the free state. See Figure 182. In others, it may be necessary to simulate the mating part interface in order to verify individual or related feature tolerances. This is done by restraining the appropriate features, such as the datum features in Figure 183. The restraining forces are those which would be exerted in the assembling or functioning of the part. The word "restrained" in a drawing note means restraint within these limitations when the amount of restraint is not specified. However, if the dimensions and tolerances are met in the free state it is usually not necessary to restrain the part unless the effect of subsequent restraining forces on the concerned features could cause other features of the part to exceed specified limits. Free state variation of nonrigid parts may be controlled as described in the following subparagraphs.

5-6.8.1 Specifying Form or Location Tolerances on Features Subject to Free State Variation. Where an individual form or location tolerance is applied to a feature in the free state, specify the maximum allowable free state variation with an appropriate feature control symbol. See Figure 182. The term FREE STATE may be placed beneath the symbol to clarify a free state requirement on a drawing containing restrained feature notes or to separate a free state requirement from associated features having restrained requirements. See Figure 183.

5-6.8.2 Specifying Form or Location Tolerances on Features to be Restrained. Where interrelated form or location tolerances are to be verified with the part in a restrained condition, select and identify the features (pilot diameter, bosses, flanges, etc.) to be used as datum surfaces. Since these surfaces may be subject to free state variation, it is necessary to specify the maximum force necessary to restrain each of them. Determine the amount of the restraining or holding forces and other requirements necessary to simulate

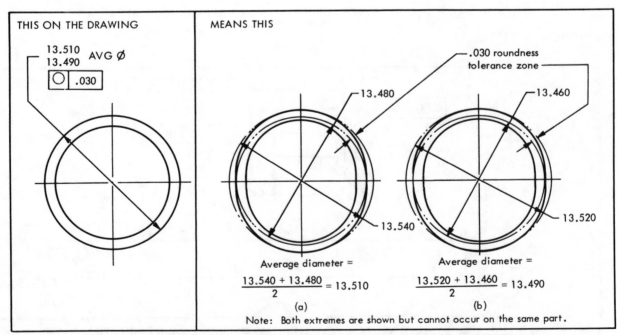

FIG. 182 SPECIFYING ROUNDNESS IN A FREE STATE WITH AVERAGE DIAMETER

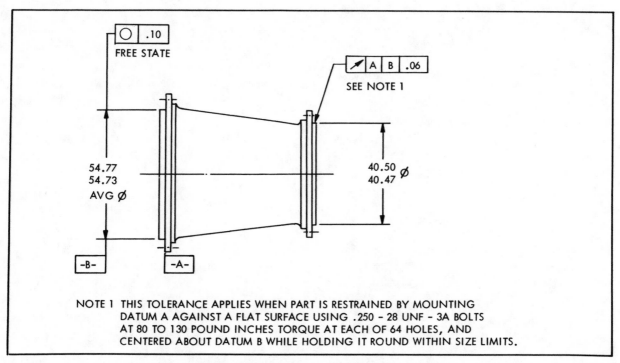

NOTE 1 THIS TOLERANCE APPLIES WHEN PART IS RESTRAINED BY MOUNTING
DATUM A AGAINST A FLAT SURFACE USING .250 – 28 UNF – 3A BOLTS
AT 80 TO 130 POUND INCHES TORQUE AT EACH OF 64 HOLES, AND
CENTERED ABOUT DATUM B WHILE HOLDING IT ROUND WITHIN SIZE LIMITS.

FIG. 183 SPECIFYING RESTRAINT FOR NONRIGID PARTS

expected assembly conditions. Specify on the drawing that if restrained to this condition, the remainder of the part or certain features thereof shall be within stated tolerances. See Figure 183.

5-6.8.3 Average Diameter. Where form control, such as roundness, is specified in a free state for a circular or cylindrical feature, the pertinent diameter is qualified with the term AVG ⌀. See Figure 182. Specifying roundness on the basis of an average diameter on a nonrigid part is necessary to ensure the actual diameter of the feature can be restrained to the desired shape at assembly. An average diameter is the average of several diametral measurements across a circular or cylindrical feature. Normally, a sufficient number of measurements, usually no less than four, are taken to assure the establishment of an average diameter. If practicable, an average diameter may be determined by a peripheral tape measurement. Note the free state roundness tolerance can be greater than the size tolerance on the diameter. Figure 182 (a) and (b), simplified by showing only two measurements, give the permissible diameters in the free state for two extreme conditions of maximum average diameter and minimum average diameter, respectively. The same method applies when the average diameter is anywhere between maximum and minimum limits.

5-7 DUAL DIMENSIONING

CAUTION: This form of dimensioning is applicable where interchangeable parts are to be manufactured in both US Customary (Inch) and SI (Metric) units of measurement. It is not recommended as a practice for implementing the transition from Inch to Metric units.

5-7.1 GENERAL. Dual dimensioning is a procedure where both U.S. customary (inch) units and SI (metric) units of measurement are shown on the same engineering drawing. This procedure permits the use of certain ISO (International Organization for Standardization) Recommendations related to drawing practices in conjunction with SI units (see 5-7.4). The International System of Units (SI) is described in the NBS (National Bureau of Standards) Special Publication 330 and in ISO/R1000.

5-7.2 INTERCHANGEABILITY. Both U.S. customary and SI dimensions on a drawing must achieve the required degree of interchangeability of parts from a drawing. Interchangeability is determined at the time of dimensional conversion by the number of decimal places retained when rounding off a converted dimension as well as to what degree the tolerance limits of the conversion are allowed to violate the limits of the original dimension. Suggested sources for conversion principles and conversion tables are:

(a) Rules for Conversion and Rounding—American National Standard Z210.1-1973, Standard Metric Practice Guide.

(b) Conversion of Toleranced Linear Dimensions—Society of Automotive Engineers J390, Dual Dimensioning Standard.

(c) National Bureau of Standards Miscellaneous Publication 286, Units of Weights and Measure, International (Metric) and U.S. Customary.

Examples herein demonstrate placement and presentation practices only and do not exemplify conversion practice.

5-7.3 IDENTIFICATION OF UNITS. Each drawing must identify which units are U.S. customary and which are SI. Two methods are illustrated to distinguish the U.S. customary unit from the SI unit, either the position or the bracket (symbol) method.

5-7.3.1 Position Method. Units may be displayed in either of two ways:

(a) Show the millimetre dimension above the inch dimension, separated by a horizontal line, or to the left of the inch dimension, separated by a virgule (slash line).

Examples:

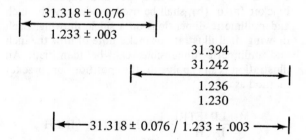

(b) Same as (a) except inch dimension above or to the left of the millimetre dimension.

5-7.3.2 Bracket Method. Units may be displayed in either of two ways:

(a) Show the millimetre dimension enclosed in square brackets (symbol [   ]). Location of the millimetre dimension is optional but shall be uniform on any drawing; that is, above or below, or to the left or the right of the inch dimension.

91

Examples:

$$[31.318 \pm 0.076]$$

$$\begin{bmatrix} 31.394 \\ 31.242 \end{bmatrix}$$

$$\longleftarrow 1.233 \pm .003 \longrightarrow | \quad \longleftarrow \begin{matrix} 1.236 \\ 1.230 \end{matrix} \longrightarrow$$

or   or

$$\longleftarrow 1.233 \pm .003 \longrightarrow | \quad \longleftarrow \begin{matrix} 1.236 \\ 1.230 \end{matrix} \longrightarrow$$

$$[31.318 \pm 0.076]$$

$$\begin{bmatrix} 31.394 \\ 31.242 \end{bmatrix}$$

or   or

$$\longleftarrow 1.233 \pm .003 \, [31.318 \pm 0.076] \longrightarrow$$

$$\longleftarrow \begin{matrix} 1.236 \\ 1.230 \end{matrix} \begin{bmatrix} 31.394 \\ 31.242 \end{bmatrix} \longrightarrow$$

or   or

$$\longleftarrow [31.318 \pm 0.076] \; 1.233 \pm .003 \longrightarrow$$

$$\longleftarrow \begin{bmatrix} 31.394 \\ 31.242 \end{bmatrix} \begin{matrix} 1.236 \\ 1.230 \end{matrix} \longrightarrow$$

(b) Same as (a), except enclose the inch dimension in square brackets instead of the millimetre dimension.

5-7.3.2.1 Computer prepared drawings may require the use of parentheses rather than brackets due to computer character limitations. In such cases, parentheses shall not be used to identify reference dimensions. See 5-1.10.5.

5-7.3.3 Only one method, position (a) or (b) or bracket (a) or (b), shall be used to identify the inch and millimetre dimensions on a single drawing. Each drawing shall illustrate or make note of how the inch and millimetre dimensions can be identified. An illustration can apply to the position or bracket method as:

$$\frac{\text{MILLIMETRE}}{\text{INCH}}; \text{MILLIMETRE} \, / \, \text{INCH}$$

or

$$\frac{[\text{MILLIMETRE}]}{\text{INCH}}; [\text{MILLIMETRE}] \, \text{INCH}$$

Note form may be used for the bracket method, as:

DIMENSIONS IN [ ] ARE MILLIMETRES

or

DIMENSIONS IN [ ] ARE INCHES

5-7.3.4 Other Units. Units other than linear dimensions and all units in notes or text on a drawing are shown in the same manner as that for dimensions, except the appropriate measurement unit symbol is specified.

Example:

$$\text{DENSITY } [32 \text{ kg/m}^3] \; 2 \text{ LB/FT}^3$$

5-7.3.5 Common Units. The callout of some unit quantities can satisfy both systems of measurement, that is, 0.006 inch per inch of taper and 0.006 millimetre per millimetre have the same rate and can be expressed as a ratio, .006:1. A note could read, TAPER .006:1.

5-7.3.6 Angles. Angles stated in degrees, minutes, and seconds, or in degrees and decimals of a degree are common to the inch and metric systems of measurement.

5-7.3.7 Nominal Designations. Designations such as nominal thread sizes, pipe sizes, and wood cross-sectional sizes, are not converted. In such cases, the origin of the drawing controls which measurement system is the basis of such designations.

5-7.4 PRACTICES APPLYING TO SI VALUES. In general, SI numerical values appear on engineering drawings in the manner prescribed by international drawing practices. The following paragraphs point out the manner in which zeros are used. Two other details are shown that apply equally to both the inch and the SI values.

5-7.4.1 Use of Zeros. A zero precedes a decimal point in an SI value of less than unity.

Example: 0.13

5-7.4.1.1 Except as indicated in 5-7.4.1.2, 5-7.4.1.3 and 5-7.4.1.4, zeros do not follow a significant digit of an SI value to the right of the decimal point. Where equal plus and minus tolerancing is used, the millimetre dimension and its tolerance may not have the same number of decimal places. In addition, if a millimetre dimension is a whole number, neither the decimal point nor a zero is shown.

5-7.4.1.2 Where unilateral tolerancing is used and either the plus or minus tolerance is nil, a single zero is shown without a plus or minus sign.

Example:

$$32 \, \begin{matrix} 0 \\ -0.02 \end{matrix} \quad \text{or} \quad 32 \, \begin{matrix} +0.02 \\ 0 \end{matrix}$$

5-7.4.1.3 Where bilateral tolerancing is used, both the plus and minus tolerance have the same number of decimal places, using zeros where necessary.

Example:

$$32 \begin{array}{c} +0.25 \\ -0.10 \end{array} \quad not \quad 32 \begin{array}{c} +0.25 \\ -0.1 \end{array}$$

5-7.4.1.4 Where limit dimensioning is used and either the maximum or minimum value has digits following a decimal point, the other value has zeros added for uniformity.

Example:

$$\begin{array}{c} 25.45 \\ 25.00 \end{array} \quad not \quad \begin{array}{c} 25.45 \\ 25 \end{array}$$

5-7.4.2 Diameter Symbol. The diameter symbol (see Figure 76) is used at all applicable places on a dual dimensioned drawing rather than the abbreviation DIA. See 5-3.3.5 for the use of the diameter symbol in a feature control symbol.

5-7.4.3 Commas. Commas shall not be used to denote thousands with either U.S. customary or SI values because it is the practice in many countries to use the comma as a decimal point sign. See 5-7.5.1. Another practice is to leave a space where a comma would otherwise be used to denote thousands. On engineering drawings, a space may be interpreted as a place where a decimal point was missing, particularly when a reproduction of the drawing was made. Therefore, spaces and commas are not permitted in numerical expressions. See following example:

Example:

32541 not 32,541 nor 32 541

5-7.5 APPLICATION. Some procedures related to the metric drawing practice and individual company drawing practices could cause difficulty in producing a dual dimensioned drawing. The following paragraphs indicate a uniform procedure that will eliminate a drafting problem (5-7.5.2) or a language translation problem (5-7.5.6) and will clarify some of the questions that seem to be frequently asked.

5-7.5.1 Decimal Point. A period is used for the decimal point sign for both U.S. customary and SI units

of measurement. (The comma, which is used as the decimal point sign in many metric or SI oriented countries, could have been used with SI units without prejudice to the principles established herein.)

5-7.5.2 Angle of Projection. Many metric or SI oriented countries use first angle projection while third angle projection is normal U.S. practice. To avoid confusion, dual dimensioned drawings shall specify the angle of projection. See American National Standard for Projections, Y14.3-1957, for appropriate symbols and Figures 184 and 185 for an application.

5-7.5.3 New Drawings. Dual dimensioning of new drawings is facilitated if all dimensions are shown in decimals except where fractions normally designate nominal sizes.

5-7.5.4 Existing Drawings with Fractional Dimensions. Existing drawings with dimensions in fractions may be dual dimensioned by adding the millimetre conversion adjacent to the fractional inch dimension. Fractions are considered the same magnitude as two place decimal inch dimensions when determining the number of places to retain in the millimetre conversion.

Examples:

$$\underline{61.91 \pm 0.4} \qquad [61.91 \pm 0.4]$$
$$\left|\leftarrow 2\frac{7}{16} \pm \frac{1}{64} \rightarrow\right|\left|\leftarrow 2\frac{7}{16} \pm \frac{1}{64} \rightarrow\right|$$

5-7.5.5 General Tolerances. General tolerances expressed as part of the drawing format or in a general note shall be dual identified.

5-7.5.6 Geometric Characteristic Symbols. Geometric characteristic symbols used for specifying tolerances of form and position shall be used on dual dimensioned drawings in accordance with the requirements of this standard.

5-7.6 DRAWING APPLICATION. Application of dual dimensioning procedures is depicted in Figures 184 and 185.

5-7.6.1 Methods of Dimensioning. The method of dimensioning shown in Figures 184 and 185 does not imply that plus and minus tolerancing is mandatory for dual dimensioned drawings.

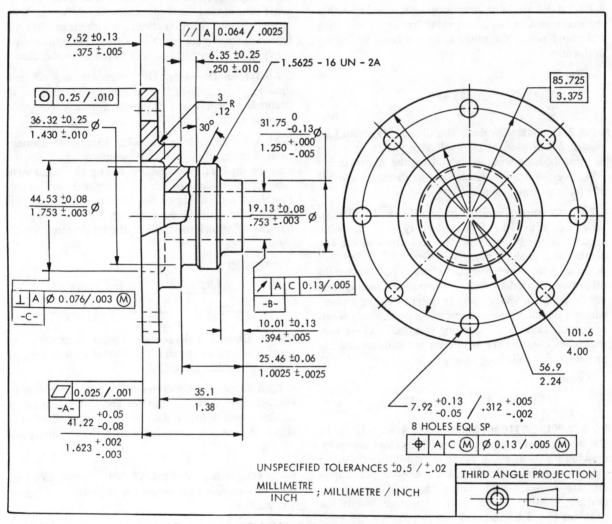

FIG. 184 POSITION METHOD

94

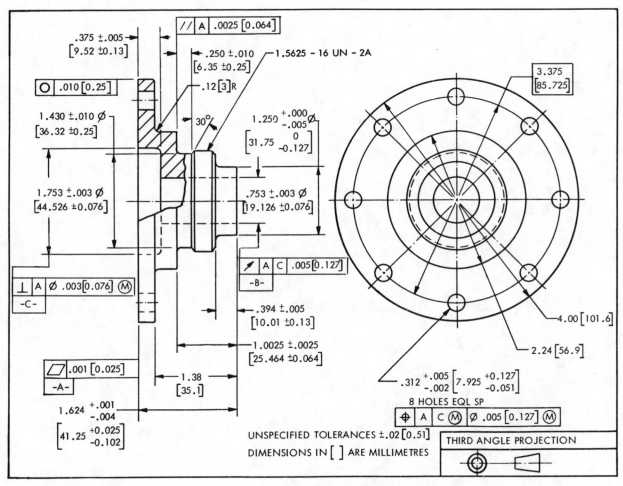

FIG. 185   BRACKET METHOD

APPENDIX   A

DIMENSIONING FOR NUMERICAL CONTROL

This Appendix is not a part of American National Standard for Dimensioning and Tolerancing, Y14.5-1973,
but is included for information purposes only.

A1 General. Acceptance of numerical control processes is being accelerated by the development of computer-aided design and computer-aided manufacturing techniques. With the rapidly changing state-of-the-art, present practices should be carefully screened and the inapplicable ones modified or deleted. These methodologies should be tailored to the particular requirements, responsibilities, and capabilities of the user. However, common guidelines can be established enabling dimensioning and tolerancing practices to be used effectively in delineating parts for both numerical control and conventional fabrication.

A2 Coordinate System. The numerical control concept is based on the system of rectangular or Cartesian coordinates wherein any position can be described in terms of distance from an origin point along either two or three mutually perpendicular axes. Two dimensional coordinates (X, Y) define points in a plane, while three dimensional coordinates (X, Y, Z) locate points in space. For effective use of this method of measurement, dimensioning practices should be used that are compatible with such a system so as to convey the precise definition to programmers, engineers, designers, draftsmen, and machine setup men.

A3 Axis Nomenclature. To establish consistency, axes on drawings are considered to intersect at an origin and should be in accordance with the following:

(a) The X axis is horizontal and parallel to the bottom edge of the drawing. It is considered the first and basic reference axis.

(b) The Y axis is vertical and perpendicular to the X axis in the plane of a drawing showing X–Y relationships, usually depicted by a plan view of the part. See upper portion of Figure A1.

(c) The Z axis is perpendicular to the plane of a drawing of X–Y relationships. For a drawing of X–Z relationships, the Z axis is vertical and perpendicular to the X axis in the plane of the drawing, usually depicted by an elevation view of the part. See lower portion of Figure A1. The intersection of the X axis and Y axis forms quadrants described in Figure A2. Axes should be aligned or coincident with a datum feature in the part view. The part view should be drawn in quadrants so that a maximum of positive values will result in programming. For example, if the

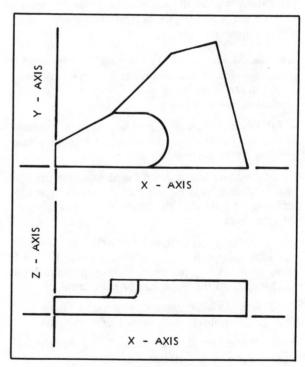

FIG. A1   PROGRAMMING AXES

97

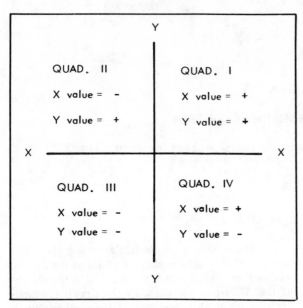

FIG. A2   MATHEMATICAL QUADRANTS

view is drawn in the positive quadrant of the intersection of X-Y axes, positive values will result. On the other hand, if the view is drawn partially in two or more quadrants, positive and negative values will result, and the potential for error is greater. This precaution is generally not necessary when programming on the computer, but helpful when programming without computer assistance.

A4 Dimensioning and Tolerancing. Recommended guidelines for dimensioning and tolerancing practices for use in defining parts for numerical control fabrication are as follows:

(a) When the basic coordinate system is established, the origin should be placed at an appropriate location on the part itself.

(b) Any number of subcoordinate systems may be used to define features of a part as long as these systems can be related to the basic coordinate system of the given part.

(c) Define part surfaces in relation to three mutually perpendicular reference planes. Establish these planes along part surfaces which parallel the machine axes if these axes can be predetermined.

(d) Use decimal dimensioning rather than fractions or a combination of decimals and fractions.

(e) Dimension the part precisely so the physical shape can be readily determined. Dimension to points on the part surfaces.

(f) Regular geometric contours such as ellipses, parabolas, hyperbolas, etc., may be defined on the drawing by mathematical formulas. The numerically controlled machinery can easily be programmed to approximate these curves by linear interpolation, that is, as series of short straight lines whose end points are close enough together to ensure meeting the required tolerances for the contour. In the case of arbitrary curves, the drawing should specify appropriate points on the curve by coordinate dimensions (or a table of coordinates). Consideration should be given to the number of points needed to define the curve, keeping in mind that the tighter the tolerance or the smaller the radius of curvature, the closer together the points should be. Such terms as "blend smoothly" and "faired curve" are not used. Curves may also be defined by other coordinates, such as polar, spherical, or cylindrical, as applicable.

(g) Changes in contour should be unambiguously defined with prime consideration for design intent.

(h) Holes in a circular pattern should preferably be located with coordinate dimensions.

(i) Express angular dimensions, where possible, relative to the X axis in degrees and decimal parts of a degree.

(j) Use plus and minus tolerances, not limit dimensions. Preferably, the tolerance should be equally divided bilaterally.

(k) Positional tolerancing, form tolerancing and datum referencing should be used where applicable. Datum features specified on the drawing in proper sequence will clearly indicate their usage for setup.

(l) Where profile tolerances are specified, the geometric boundary should be equally disposed bilaterally along the true profile. Avoid profile tolerances applied unilaterally along the true profile. Include no less than four defined points along the profile.

(m) Tolerances are specified only on the basis of actual design requirements. The accuracy capability of numerically controlled equipment is not basis for specifying more restrictive tolerances than functionally required.

A5 Computer Programming Language. The job of the programmer is greatly simplified if the designer is familiar with the particular program language to be used to reproduce the part. This knowledge will help him to anticipate information the programmer needs to produce the tape and will smooth the flow of information from designer to numerically controlled machine tool.

APPENDIX B

ADVANTAGES OF POSITIONAL TOLERANCING

This Appendix is not a part of American National Standard for Dimensioning and Tolerancing, Y14.5-1973,
but is included for information purposes only.

B1 General. The purpose of this appendix is to explain the advantages of using positional tolerancing as described in Subsection 5-5 of this standard.

B2 Advantage of the Cylindrical Tolerance Zone. With traditional coordinate plus and minus tolerancing, a square, rectangular, or wedge-shaped tolerance zone results. See Figures B1, B2 and B3. The resultant tolerance zone for the feature dimensioned in Figure B1 (a) is a 0.010 square as shown in Figure B2. Since the actual position of the feature may be anywhere within the square, the maximum allowable variation from the desired (or normal) position occurs at 45 degrees from the horizontal or vertical direction and is equal to about 1.4 times the specified tolerance. See Figure B4.

B2.1 Actually, the maximum allowable tolerance indicated for the 45 degree direction in Figure B4 is permissible in any direction without detrimental effect on the assembly of the mating parts. This fact is used to advantage in positional tolerancing. The positional tolerance zone corresponding to the 0.010 square zone of Figure B4 is a cylinder of 0.014 diameter which provides an increase of 57 percent in target area for location of the feature. See Figure B5. Such increase in tolerance often results in reduction in cost of manufacture.

B3 Practical Gaging. For quantity inspection, the use of position or receiver gages (with round pins) is one of the most practical kinds of gaging for inspection of patterns of holes. Such round pin gaging can properly inspect parts only where holes have a tolerance zone that is cylindrical.

B4 Uniform Interpretation and Simpler Analysis. Coordinate plus and minus tolerancing for complex hole patterns is subject to more than one interpretation and therefore difficult to analyze. Where positional tolerancing is used, only one interpretation is possible, making it easy to determine permissible tolerance values and clearance hole sizes to assure assembly without interference.

B4.1 Figures B7, B8 and B9 are three tolerance zone patterns that are acceptable interpretations of the coordinate plus and minus tolerances specified in Figure B6. Other variations of these tolerance zones also would meet the requirements of Figure B6.

B4.2 Figure B10 shows the same part as Figure B6, but specified in accordance with positional tolerancing principles. The resulting tolerance zone pattern is

99

shown in Figure B11. Although location of the four tolerance zones for the interrelationship of holes (as a group) may vary from that shown, only one interpretation is possible as to the size and shape of the tolerance zones and the relationship between their axes.

B5 Advantage of the MMC Concept. Parts are generally toleranced so they will assemble when mating features are at maximum material condition (MMC), taking into consideration the variation in position of the features. For example, in a bolt pattern, clearance holes are provided such that parts will assemble when bolts are at their largest diameter, the holes at their smallest diameter and the center distance between holes are at extremes of the permissible tolerance. If in manufacturing, holes are made larger than MMC size, parts will assemble even though hole locations are outside the specified tolerance. Additional tolerance on location can be permitted when features depart from their MMC size. With coordinate plus and minus tolerancing, the drawing does not provide recognition of this additional tolerance; therefore usable parts may be rejected. With positional tolerancing, application of the MMC symbol allows additional tolerance of position as the feature departs from MMC. The following analysis of a simple two-hole pattern using Figures B12 through B17 illustrates the MMC concept:

(a) Figure B12 shows the drawing requirements for location and size of two holes. If the holes are exactly 0.500 diameter (the maximum material condition or the smallest size hole permitted by the drawing specification) and are centered exactly 2.000 apart, they will theoretically receive a gage consisting of two round pins fixed in a plate if the pins are centered 2.000 apart and are 0.500 diameter.

(b) However, the limits specified on the drawing require the center distance between holes to be from 1.993 to 2.007, if the hole is at minimum limit of size. The pins in the gage would have to be 0.007 smaller, or 0.493 diameter, to enter the holes in their extreme positions, as shown in Figure B13.

(c) If the holes are exactly 0.500 diameter but located at the maximum permissible center distance, the 0.493 diameter gage pins would contact the inner sides of the holes since the distance between the inner sides of the gage pins is 1.507. See Figure B14.

(d) If the holes are exactly 0.500 diameter but located at the minimum permissible center distance, the gage pins would contact the outer sides of the holes if the distance between the outer sides of the gage pins is 2.493. See Figure B15. Neglecting gage-maker's tolerances, the gage pins would have to be 0.493 diameter and the centers located exactly 2.000 apart. Thus, the holes in the part, if they are 0.500 diameter, will fit the gage pins if located within the limit specified on the drawing.

(e) If the holes are at maximum size, 0.505 diameter permitted by the drawing, they will be accepted by the gage if the inner sides of the holes contact the inner sides of the gage pins, which means the holes will have a center distance of 2.012. See Figure B16.

(f) These holes will also be accepted by the gage when their outer sides contact the outer sides of the gage pins or when at a center distance of 1.988. See Figure B17.

B6 The Advantage of Zero Positional Tolerance at MMC. In the application of positional tolerancing shown in preceding paragraphs and in Subsection 5-5, a positional tolerance of some magnitude is always specified and is calculated to allow for the condition where the mating features are at their extreme limits of position and size tolerances. However, rejection of usable parts can occur where these features are actually located on or close to their true positions but produced to a size smaller than the specified minimum (outside of limits). The concept of zero tolerance at MMC, where specified, permits acceptance of parts over the widest possible tolerance range.

B6.1 Application. Figures B18 and B19 will illustrate the concept. Figure B18 shows a drawing for one of two identical plates to be fastened together with four 0.500 maximum diameter bolts. Using conventional positional tolerancing, 0.531 diameter clearance holes are selected with a tolerance as shown. The required positional tolerance is found by the formula given in Appendix C, paragraph C3:

$$T = H - F$$
$$= .529 - .500 = .029 \text{ diameter}$$

Note that if the clearance holes were located exactly at true position, the parts would still assemble with clearance holes as small as 0.500 diameter (or slightly larger). However, the drawing calls for rejection of all parts having clearance holes smaller than 0.529 diameter. Figure B19 shows a drawing of the same part toleranced according to the concept of zero positional tolerance at MMC. Note that the clearance holes at minimum diameter will just pass a 0.500 diameter bolt, and at maximum diameter are equal to the size allowed in Figure B18. Although the positional tolerance is specified as zero at MMC, the positional tolerance allowed is in direct proportion to the actual clearance hole size as shown by the following tabulation:

If the clearance hole diameter is:	the positional tolerance diameter allowed is:
.500	.000
.501	.001
.502	.002
↑	↑
etc.	etc.
↓	↓
.539	.039

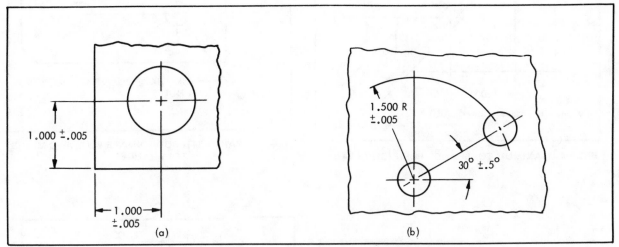

FIG. B1   COORDINATE PLUS AND MINUS TOLERANCING

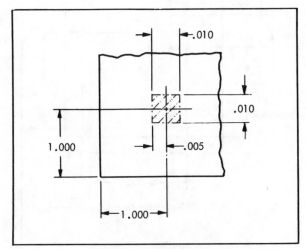

FIG. B2   SQUARE TOLERANCE ZONE FOR
FIGURE B1(a)

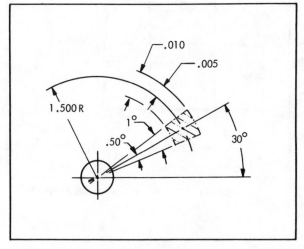

FIG. B3   WEDGE-SHAPED TOLERANCE ZONE FOR
FIGURE B1(b)

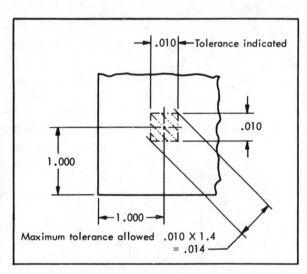

FIG. B4   MAXIMUM TOLERANCE FOR FIGURE B1(a)

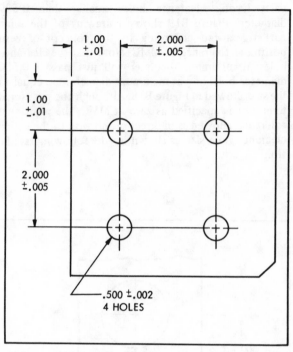

FIG. B6   PART WITH COORDINATE PLUS AND MINUS
TOLERANCES

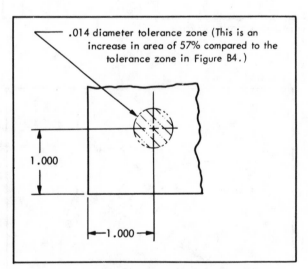

FIG. B5   POSITIONAL TOLERANCE ZONE

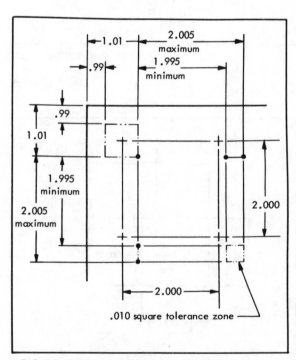

FIG. B7   ONE INTERPRETATION FOR FIGURE B6

102

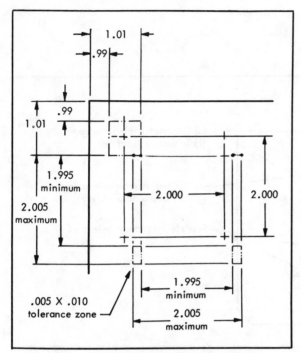

FIG. B8 A SECOND INTERPRETATION FOR
FIGURE B6

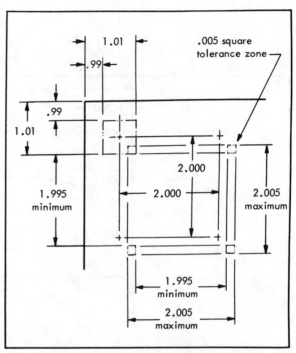

FIG. B9 A THIRD INTERPRETATION FOR
FIGURE B6

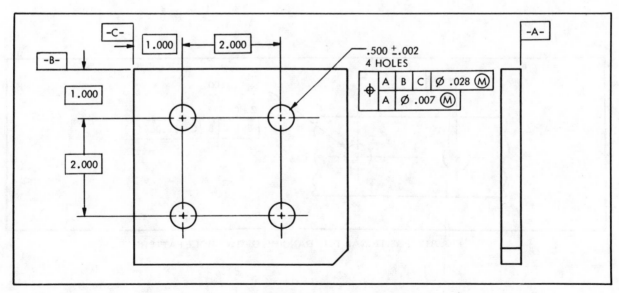

FIG. B10 POSITIONAL TOLERANCING OF PART SHOWN IN FIGURE B6

103

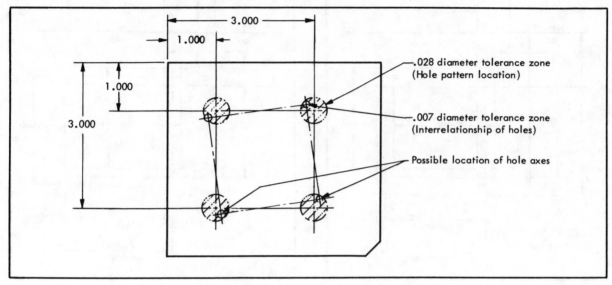

FIG. B11   TOLERANCE ZONES FOR FIGURE B10

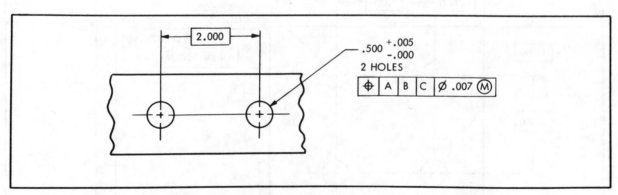

FIG. B12   POSITIONAL TOLERANCING OF TWO-HOLE PATTERN

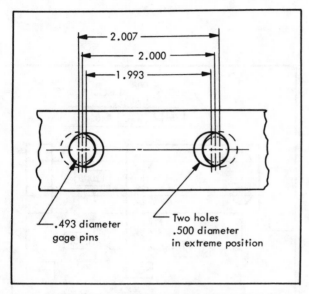

FIG. B13   EXTREME POSITIONS AT MMC

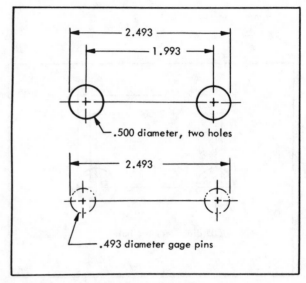

FIG. B14   MAXIMUM CENTER DISTANCE, MMC

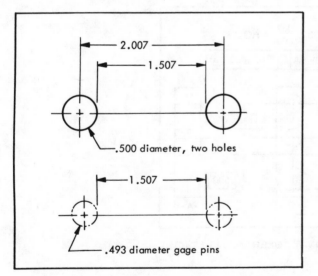

FIG. B15   MINIMUM CENTER DISTANCE, MMC

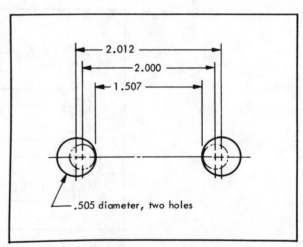

FIG. B16   MAXIMUM CENTER DISTANCE, HOLES AT
MAXIMUM DIAMETER

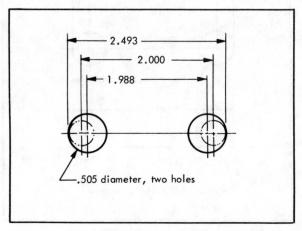

FIG. B17   MINIMUM CENTER DISTANCE, HOLES AT
MAXIMUM DIAMETER

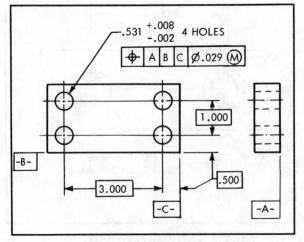

FIG. B18   CONVENTIONAL POSITIONAL
TOLERANCING AT MMC

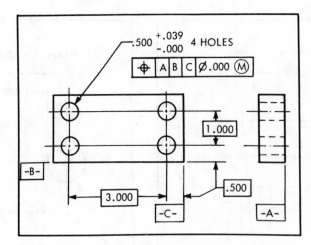

FIG. B19   ZERO POSITIONAL TOLERANCING AT MMC

APPENDIX C

FORMULAS FOR POSITIONAL TOLERANCING

This Appendix is not a part of American National Standard for Dimensioning and Tolerancing, Y14.5-1973,
but is included for information purposes only.

C1 General. The purpose of this appendix is to present formulas for determining the required positional tolerances or the required sizes of mating features to ensure that parts will assemble. The formulas are valid for all types of features or patterns of features and will give a "no interference, no clearance" fit when features are at maximum material condition with their locations in the extreme of positional tolerance.

C2 Symbols. Formulas given herein use the three basic symbols listed below:

F = Maximum diameter of fastener
H = Minimum diameter of clearance hole
T = Positional tolerance diameter

Subscripts are used when more than one size feature or tolerance is involved.

C3 Floating Fastener Case. Where two or more parts are assembled with fasteners such as bolts and nuts, and all parts have clearance holes for the bolts, it is termed the floating fastener case. See Figure C1.

Where the fasteners are of the same diameter, and it is desired to use the same clearance hole diameters and the same positional tolerances for the parts to be assembled, the following formula applies:

$$H = F + T \text{ and } T = H - F$$

Example: Given that the fasteners in Figure C1 are 0.138 diameter maximum and the clearance holes are 0.164 diameter minimum. Find the required positional tolerance:

$T = 0.164 - 0.138 = 0.026$ diameter for each part

If desired, the positional tolerance may be distributed unequally between parts. To do this, T is separated into T_1 and T_2 such that:

$$T = \frac{T_1 + T_2}{2}$$

In the above example T_1 could be 0.020, then $T_2 = 0.032$ since $\dfrac{0.032 + 0.020}{2} = 0.026$

Note where three or more parts are considered for unequal tolerance distribution, the above formula must be satisfied by any combination of two parts.

The general formula for the floating fastener case where two mating parts have different positional tolerances and different size clearance holes is:

$$H_1 + H_2 = 2F + T_1 + T_2$$

C4 Fixed Fastener Case. Where one of the parts to be assembled has restrained fasteners such as screws in tapped holes or studs, it is termed the fixed fastener case. See Figure C2.

Where the fasteners are of the same diameter and it is desired to use the same positional tolerances in the parts to be assembled, the following formula applies:

$$H = F + 2T \text{ and } T = \frac{H - F}{2}$$

Note that the allowable positional tolerance is half that for the comparable floating fastener case.

Example: Given, the fasteners in Figure C2 have a maximum diameter of 0.138 and the clearance holes have a minimum diameter of 0.164. Find the required positional tolerance:

$$T = \frac{0.164 - 0.138}{2} = 0.013 \text{ diameter for each part}$$

If it is desired that the part with tapped holes has a larger positional tolerance than the part with clearance holes, T can be separated into T_1 and T_2 in any appropriate manner such that:

$$T = \frac{T_1 + T_2}{2}$$

For example T_1 could be 0.010, then T_2 would be 0.016.

The general formula for the fixed fastener case where two mating parts have different positional tolerances is:

$$H = F + T_1 + T_2$$

C5 Coaxial Features. The formula previously given for the floating fastener case also applies to mating parts having two coaxial features where one of these features is a datum for the other. See Figure C3.

Where it is desired to divide the available tolerance unequally between the parts, the following formula is useful:

$$H_1 + H_2 = F_1 + F_2 + T_1 + T_2$$

(This formula is valid only for simple two feature parts as shown here.)

Example: Given the information shown in Figure C3, solve for T_1 and T_2.

$$H_1 + H_2 = F_1 + F_2 + T_1 + T_2$$
$$T_1 + T_2 = (H_1 + H_2) - (F_1 + F_2)$$
$$= (1.002 + 0.501) - (1.000 + 0.500)$$
$$= 0.003 \text{ total available tolerance}$$

If $\quad T_1 = 0.002$

$\qquad T_2 = 0.001$

C6 Provision for Out-of-Squareness. The formulas do not provide sufficient fastener clearance for the fixed fastener case when threaded holes or holes for tight fitting members, such as dowels, are out of square. To provide for this condition, the "projected tolerance zone" method of positional tolerancing should be applied to threaded holes or tight fitting holes. See 5-5.7.

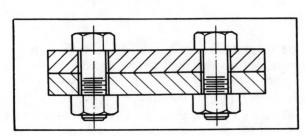

FIGURE C1 FLOATING FASTENER

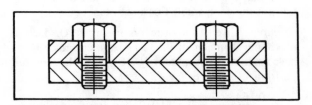

FIGURE C2 FIXED FASTENER

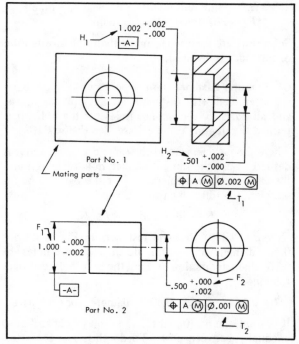

FIG. C3 COAXIAL FEATURES

APPENDIX D

FORM, PROPORTION AND COMPARISON OF SYMBOLS

This Appendix is not a part of American National Standard for Dimensioning and Tolerancing, Y14.5-1973, but is included for information purposes only.

D1 General. The purpose of this appendix is to present the recommended form and proportion for geometric characteristic symbols. This information should assist manufacturers of drawing templates in producing aids to achieve uniform results. While it is expected that most professional draftsmen will use commercial templates, this appendix will undoubtedly assist others to obtain uniform results without such aids.

D2 Proportions. Figure D1 illustrates the preferred proportions of the various geometric characteristic symbols, based primarily on modules of 50 percent and 75 percent of the basic frame height. The symbols are grouped to illustrate similarities in the elements of their construction.

D3 Comparison. Figure D2 provides a comparison of the symbols adopted by this standard with those of ISO, British and Canadian standards.

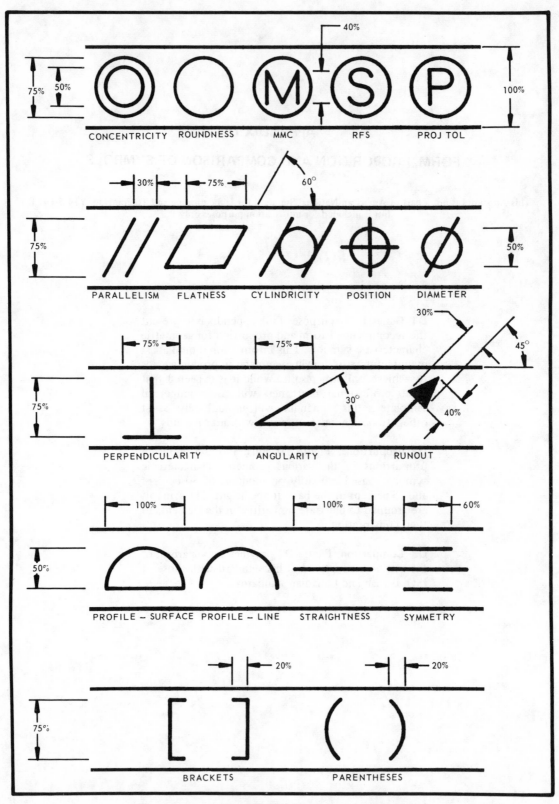

FIG. D1   FORM AND PROPORTION OF SYMBOLS

Characteristics	American ANSI Y14.5	British BS 308	Canadian CSA B78.2	International ISO R1101
Straightness	—	Same	Same	Same
Flatness	▱	Same	Same	Same
Roundness (Circularity)	○	Same	Same	Same
Cylindricity	⌭	Same	Same	Same
Profile of a Line	⌒	Same	Same	Same
Profile of a Surface	⌓	Same	Same	Same
Parallelism	//	Same	Same	Same
Perpendicularity (Squareness)	⊥	Same	Same	Same
Angularity	∠	Same	Same	Same
Position	⌖	Same	Same	Same
Concentricity (Coaxiality)	◎	Same	Same	Same
Symmetry	⛛	Same	Same	Same
Maximum Material Condition	Ⓜ	Same	Same	Same
Diameter	∅	Same	Same	Same
Circular Runout	↗	Same	Same	Same
Total Runout	↗ ✱	None	None	None
Datum Identification	-A-	A or ⌐	-A- or ⌐	A or ⌐
Reference Dimension	(5.000)	(127)	(5.000)	(127)
Basic Dimension	5.000	127	5.000	127
Regardless of Feature Size	Ⓢ	None	None	None
Projected Tolerance Zone	Ⓟ	None	Ⓟ	None
Datum Target	A/1	A/1	A/1	None
Part Symmetry	None	―‖―·―‖―	―‖―·―‖―	―‖―·―‖―
Shape of the tolerance zone	Zone is total width. ∅ specified where zone is circular or cylindrical.	Zone is a total width in direction of leader arrow. ∅ specified where zone is circular or cylindrical.	Zone shape evident from chacteristic being controlled.	Zone is a total width in direction of leader arrow. ∅ specified where zone is circular or cylindrical.
Sequence within the feature control symbol	⌖ A B C ∅.02 Ⓜ or ⌖ ∅ .02 Ⓜ A B C	⌖ ∅ 0.5 Ⓜ A B C	⌖ .02 Ⓜ A B C	⌖ ∅ 0.5 Ⓜ A B C

✱ "TOTAL" specified under the feature control symbol.

FIG. D2 COMPARISON OF SYMBOLS

Index

G

H

E

I

F

K

American National Standards of Particular Interest to Designers, Architects and Draftsmen

TITLE OF STANDARD

Binders for holding Standards are available. A complete list of American National Standards published by The American Society of Mechanical Engineers is available upon request.